A COMPREHENSIVE CHRONICLE OF MOTORCYCLES THROUGHOUT THE WORLD

JOHN CARROLL

PUBLISHED BY
SALAMANDER BOOKS LIMITED
LONDON

A Salamander Book

Published by Salamander Books Ltd.
129–137 York Way
London N7 9LG
United Kingdom

9 8 7 6 5 4 3 2 1

CREDITS
Editor: Dennis Cove
Designer: Mark Holt
Colour and monochrome reproductions:
    Pixel Tech., Singapore
Filmset: SX Composing DTP, England

Printed and bound in Great Britain by
Butler & Tanner Ltd, Frome and London

Additional captions:
1 Dash of 1994 Indian Century Chief
2 Yamaha YZF600R Thundercat
4 1991 Triumph Daytona

CONTENTS

INTRODUCTION

The motorcycle industry in the widest sense is massive both because it is worldwide and because of the sheer diversity of machines manufactured. This diversity is partially explained by the fact that motorcycles are produced all around the world; in many countries, particularly developing ones, the motorcycle still fulfils a role as cheap and basic transportation while in others the motorcycle has become a complete luxury and is used purely for recreation. The requirements of each of these two user groups are fundamentally different, as one requires completely basic, simple and cheap machines while the other is more concerned with speed and style. This has, on occasions, led to outmoded machines from one country being manufactured in another – for example, obsolete British and Japanese machines are still manufactured in India and China respectively. Add to this the fact that motorcycles have become more specialized as they become more developed so that a machine built for arduous cross-country use will now bear little resemblance to a roadgoing sports bike. Fashion is also a major factor. Some factories build machines that look like racing bikes while others build motorcycles that look like choppers and some, such as the big Japanese manufacturers, build both. Bearing all this in mind it is possible to get an idea of the size and diversity of the motorcycle industry.

The motorcycle is approximately a century old and in that time numerous companies have come and gone; in the earliest days of motorcycling there were literally hundreds of companies (many of which already made bicycles) manufacturing the fledgling means of transport; few survived more than a few years. The ones that did were invariably those with the machines that worked but even many of these companies were wiped out by economic catastrophes such as the Great Depression. Later events such as World War II had a similar effect too. The general trend has been for an overall reduction in the numbers worldwide, although there are exceptions; many of the Italian manufacturers were founded and flourished in the aftermath of World War II, for example. The position toward the end of the 1990s is that there are four strong Japanese manufacturers, a major American manufacturer and a few European manufacturers.

Over the course of the century the number of American manufacturers has dwindled from in excess of 200 to just one major producer, Harley–Davidson. Their

INTRODUCTION

last domestic competitor, Indian, went out of business in 1953. Harley–Davidson's own fortunes experienced an upturn in the early 1980s when a change of management and a swing in fashion gave the company a very necessary shot in the arm. It has been a similar story in Great Britain where the motorcycle industry changed from being the world leader to almost nothing, before a new Triumph company began to produce world class motorcycles again. The American story was repeated in Germany where as a result of World War II, the Iron Curtain and economic trends, the number of German motorcycle manufacturers contracted dramatically, leaving BMW almost in isolation. In Italy, a country that has produced machines as diverse as the MV Agusta and the Lambretta, several of the companies founded in the postwar years that produced sporting motorcycles struggled financially and a number of the surviving companies are now part of Cagiva.

In many cases it is the Japanese factories that meet the need for basic transport in developing nations, and in some cases they have built production plants in these countries. The Honda Cub, a small capacity step-through moped is popular in such countries and is believed to be the most mass produced two-wheeler of all time, selling well all around the world. The Honda Cub is one of the many machines profiled within this encyclopedia. It covers a comprehensive selection of the world's manufacturers and a representative sample of the machines they made or still make. Motorcycles are profiled individually with a specification panel and through the listings it is possible to trace the development of the motorcycle from its origins as a bicycle fitted with a tiny engine to a huge and highly technical, superbly engineered machine. It is also possible to see the rise and fall of the renowned British bike industry and the subsequent rise of the Japanese bike industry to take its place, while contemporary Harley–Davidsons, Triumphs and Ducatis illustrate the renaissance of the non-Japanese manufacturers, a phenomenon of the 1990s.

John Carroll

A

ACE

The Ace company was founded by William Henderson in Philadelphia, USA, in 1919, after he had sold on his previous concern, Henderson, to Ignatz Schwinn. The Ace company produced in-line four-cylinder motorcycles to Henderson's design. The first machines from Ace had a capacity of 1168cc (71.24cu. in.), which was subsequently increased. When Henderson was killed in an accident in 1922, while testing a new motorcycle, the company continued in business with Arthur Lemon as designer. The engines were of inlet-over-exhaust design and were fitted with alloy pistons. The company experienced some financial difficulties during the 1920s and was eventually acquired by Indian in 1927.

1928 INDIAN ACE

Indian moved production to Springfield, Massachusetts, and renamed the machine

BELOW: The Ace Motorcycle company concentrated on the production of in-line four-cylinder machines such as this 1928 model.

the Indian Ace. After 1929 the in-line fours were renamed Indian Fours and Ace were consigned to the history books. The first Indian Aces differed only from Ace's product in that they were painted in Indian colors and had smaller diameter wheels.

SPECIFICATION
Country of origin: USA
Capacity: 1265cc (77.16cu. in.)
Engine cycle: 4-stroke
Number of cylinders: 4
Top speed: 75mph (120 kph) (Estimate)
Power: 30bhp (Estimate)
Transmission: 3-speed
Frame: Rigid steel

ADLER

Based in Frankfurt, Germany, Adler produced motorcycles with engines of their own manufacture for the first decade of the 20th century. From around 1910 until the outbreak of World War II in 1939 the company concentrated on the production of other items, including typewriters. In 1949 production of motorcycles was taken up once more, with the manufacture of a lightweight two-stroke. From this, Adler progressed to produce and market, for a short period, a range of slightly higher displacement. However, the company's return to

ABOVE: This 1956 Adler scooter was made just before the company became part of Grundig.

motorcycle production was short-lived and ended completely when Adler became part of Grundig, when once again production was switched back solely to typewriters.

1953 ADLER MB250

This machine was a development of Adler's successful two-stroke 195cc (11.89cu. in.) twin MB200 of 1951. The increased capacity was achieved by both bore and stroke being 54mm. The engine, fitted with a 180° crankshaft, was noted for its reliability. The clutch was on the crankshaft end and primary drive to the gearbox was by means of helical gears. The MB250 had a leading link front fork assembly, a plunger-style frame and the final drive chain was fully enclosed. Despite this obvious quality the difficulties faced by German motorcycle manufacturers

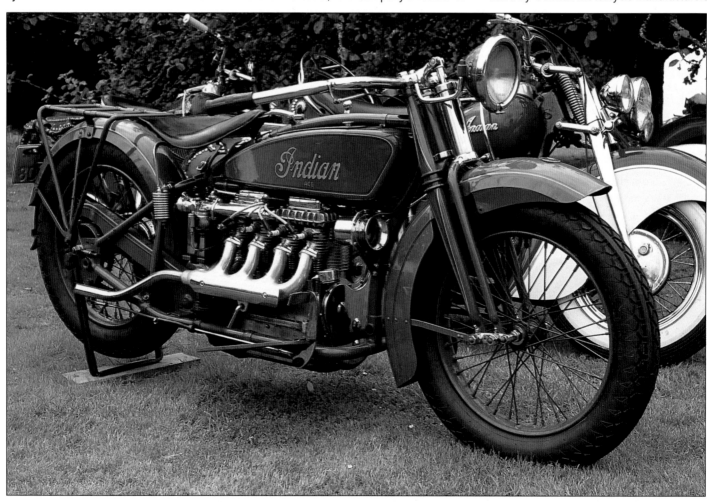

toward the end of the 1950s meant that production of the MB250 was short-lived.

SPECIFICATION
Country of origin: GERMANY
Capacity: 247cc (15.06cu. in.)
Engine cycle: 2-stroke
Number of cylinders: 2
Top speed: 65mph (104kph) (Estimate)
Power: 18bhp @ 6000rpm
Transmission: 4-speed
Frame: Tubular steel plunger

ABOVE: A 1953 Adler MB250, a two-stroke. Production of this model was short-lived.

AERMACCHI

Aermacchi was founded by Giulio Macchi in 1912, in Italy, with the intention of building aircraft. This the company did, with some success, producing both civilian and military aircraft until the end of World War II. Prohibited from building aircraft after the cessation of hostilities, the company went into the production of a three-wheeler truck. The move towards motorcycle manufacture came with the arrival of a new designer, Ing Lino Tonti. His first design, in 1950, was unorthodox, being an open-frame lightweight with a 123cc (7.50cu. in.) two-stroke engine with a single horizontal cylinder. This motorcycle was capable of 47mph (75kph) and was built with a three-speed transmission. Subsequently, a 246cc (15cu. in.) variant was produced through the use of two 123cc (7.50cu. in.) engines siamesed together.

The company then had its first forays into competitive motorcycling, with entries in the International Six Days' Trial. Tonti also designed some aerodynamic, enclosed motorcycles for successful attempts on various speed records. Tonti left soon afterward and went to work for a rival company and his place at Aermacchi was taken by Ing Alfredo Bianchi.

The first motorcycle designed by Bianchi was known as the Chimera and its futuristic design was highly acclaimed at the 1956 Milan Show, although subsequently it proved to be a costly flop in terms of sales. However, something was salvaged from the experience because, undressed of its body-

work, the bike became the basis for Aermacchi's sports and racing machines.

Harley–Davidson acquired a 50 percent stake in Aermacchi in 1960. The American company was intent on supporting its heavyweight motorcycle with lightweight European machines. The racing side of Aermacchi's motorcycle business prospered and riders of Aermacchi machines started to appear on winners' podiums for road and circuit racing as well as for motocross. The range of road bikes was not inconsiderable, with eight varying models offered in 1960, with capacities from 125cc (7.62cu. in.) up to 250cc (15.25cu. in.). A great deal of Aermacchi's production went to the USA and the whole range was refined throughout the 1960s.

In 1972 Harley–Davidson acquired sufficient shares in the company to take control. They then redesigned the range, dropping the name 'Aermacchi' from the tank and replacing it with 'AMF Harley–Davidson'. AMF put the company into receivership in May 1978 and soon afterward it was acquired by Cagiva.

1964 AERMACCHI 250

The Ala D'Oro and Ala Verde were refined for 1964 and Aermacchi's 250cc engine was redesigned so as to become a short-stroke

BELOW: The 1964 Aermacchi 250 was a single-cylinder machine with a sporting pedigree.

with bore and stroke measurements of 72mm and 61mm respectively and its camshaft would from then on be mounted in needle roller bearings. The roadgoing Aermacchis benefited from racing exploits and development, although the company's catalog clearly sold both race bikes and roadsters as separate machines.

SPECIFICATION
Country of origin: ITALY
Capacity: 247cc (15.06cu. in.)
Engine cycle: 4-stroke
Number of cylinders: 1
Top speed: 85mph (140kph)
Power: 28bhp @ 9500rpm
Transmission: 4-speed
Frame: Tubular steel spine

1964 AERMACCHI 350 ALA D'ORO

The 350cc Ala D'Oro was a direct development from the 250cc version and it was developed with racing success in mind. In 1964 only works-supported riders used the 350 bike but later they were made more widely available and production of derived racers continued until the early part of the 1970s.

SPECIFICATION
Country of origin: ITALY
Capacity: 344cc (20.98cu. in.)
Engine cycle: 4-stroke
Number of cylinders: 1
Top speed: 105mph (170kph)
Power: 28.5bhp @ 8000rpm
Transmission: 5-speed
Frame: Tubular steel spine

BELOW: Racing was one of the facets of motorcycling at which the Aermacchi factory excelled. The 350, seen here in race trim, was developed especially for competition use.

Above: This 1934 AJS export Model 34/2 was a 998cc (60.87cu. in.) four-stroke V-twin.

SPECIFICATION
Country of origin: GREAT BRITAIN
Capacity: 998cc (60.87cu. in.)
Engine cycle: 4-stroke
Number of cylinders: 2
Top speed: n/a
Power: n/a
Transmission: 4-speed
Frame: Steel duplex

AJS

Like so many British manufacturers, AJS were originally based in the English Midlands. Joe Stevens owned the Stevens' Screw Company which he had founded in 1897. His four sons – George, Jack, Joe and Harry – built the first AJS motorcycle there in 1909. It featured a 298cc (18.17cu in) side-valve engine of their own design. In the following years the company produced one of the first British overhead-valve machines. which displaced 348cc (21.22cu. in.). 1925 saw the production of a 498cc (30.37cu. in.) overhead-valve engine.

In 1931 the Stevens brothers had to sell AJS due to the disastrous effects of diversifying from the motorcycle business in the Depression years of the 1930s. The company was bought by the Collier brothers, proprietors of Matchless motorcycles. Matchless was based in Plumstead in London and AJS was moved to the same site. In the years afterward both marques had many common components.

1915 AJS MODEL D

The Model D was a bike typical of its time: a V-twin that was introduced in the years prior to World War I and gradually refined, its displacement increased. The Model D was the first of a range of V-twins from AJS which only ended with the outbreak of World War II and the company's move to singles for the Allied forces.

SPECIFICATION
Country of origin: GREAT BRITAIN
Capacity: 748cc (45.62cu. in.)
Engine cycle: 4-stroke
Number of cylinders: 2
Top speed: n/a
Power: n/a
Transmission: 3-speed
Frame: Steel diamond

Above: Many AJS motorcycles, such as this 1915 machine, were used to pull sidecars.

1934 AJS MODEL 34/2

In 1933 AJS adopted a policy of giving their motorcycles a model number, of which the first two digits were the last two of the year in which it was manufactured, so that the Model 34/2 could only be a 1934 Model 2. It featured a magneto and a single carburetor. The latter component was mounted between the cylinders. Also in the range for sale in that year was a smaller capacity 498cc (30.37cu. in.) transversely-mounted V-twin, which is generally perceived as the Stevens' attempt to increase sales through selling unusual technology. The drive from engine to gearbox was by means of a rotating shaft while final drive was by the more traditional chain.

1947 AJS 'PORCUPINE'

The AJS 'Porcupine' was originally designed as a supercharged racer but prior to the resumption of postwar racing, supercharging was banned. There was not time to redesign the engine for atmospheric induction so the castings were used featuring horizontal cylinders and the finned cylinder heads that gave the bike its nickname. The bike was introduced for 1947 and refined for several seasons. Rider Les Graham had considerable racing success aboard the 'Porcupine' including a win in the 1950 Swiss Grand Prix.

Below: The AJS Porcupine was primarily designed and intended for supercharging. This method of induction was banned after World War II so it had to be raced with carburetors.

SPECIFICATION
Country of origin: GREAT BRITAIN
Capacity: 498cc (30.37cu. in.)
Engine cycle: 4-stroke
Number of cylinders: 2
Top speed: n/a
Power: 40bhp @ 7600rpm
Transmission: 4-speed
Frame: Steel duplex loop

AMAZONAS

A relatively short-lived Brazilian marque, the Amazonas was a utility motorcycle propelled by a flat-four Volkswagen car engine. The motorcycle also utilized some other VW components such as shock absorbers and disc brakes. Production of these machines, which were assembled in Manaus, Brazil, lasted for only 12 years, between 1978 and 1990.

1986 AME 1600

The AME 1600 was used in conjunction with a sidecar from Koch Motorrad in West Germany. It was not the most aesthetically pleasing motorcycle due to its heavy reliance on VW car components.

SPECIFICATION
Country of origin: BRAZIL
Capacity: 1581cc (96.44cu. in.)
Engine cycle: 4-stroke
Number of cylinders: 4

Top speed: 108mph (175kph)
Power: n/a
Transmission: 4-speed plus reverse
Frame: Duplex tubular cradle

APRILIA

This Italian company is a relative newcomer, being founded as recently as 1968. It started out producing small capacity motorcycles for trials and for motocross. During the 1980s the company began campaigning Grand Prix motorcycles in 125cc and 250cc classes. Initially the race bikes used Rotax engines but latterly the company used its own design of engines. The first 250 World Championship won for Aprilia was in 1994 when Max Biaggi triumphed. The co-operation with Austrian engine manufacturers Rotax also led to the appearance of motorcycles such as the Pegaso, which was a 652cc (39.77cu. in.) machine, for both on- and off-road riding.

1995 APRILIA PEGASO 650

The Pegaso is clearly derived from the Paris-Dakar style of on- and off-road bikes that became popular with the annual running of that marathon race. The Pegaso is a high-tech example featuring rising-rate rear suspension and upside-down forks. The engine is liquid-cooled and has a five-valve cylinder head; valve actuation is by chain-driven overhead camshafts.

SPECIFICATION
Country of origin: ITALY
Capacity: 652cc (39.77cu. in.)
Engine cycle: 4-stroke
Number of cylinders: 1
Top speed: 108mph (175kph)
Power: 49bhp @ 7000rpm
Transmission: 5-speed
Frame: Tubular alloy twin cradle

1995 APRILIA MOTO 6.5

The 6.5 in this model's designation refers to its capacity. The bike was designed by Philippe Starck and as such is somewhat avant-garde, even for the 1990s. It has been acknowledged as deliberately mould-breaking but remains conventional to ride. Components such as the radiator and exhaust pipe are cleverly designed into the overall shape of the motorcycle so as to remain unobtrusive.

SPECIFICATION
Country of origin: ITALY
Capacity: 649cc (39.58cu. in.)
Engine cycle: 4-stroke
Number of cylinders: 1
Top speed: 95mph (152kph)
Power: 45bhp (Estimate)
Transmission: 5-speed
Frame: Tubular steel twin cradle

BELOW: The Aprilia Moto 6.5 of 1995 was a radical design even by mid-1990s standards.

ABOVE: AJS and Matchless amalgamated to
form AMC – Associated Motorcycles Company
– in 1937. From then on the marques were
simply badge-engineered variants of each
other. This is a 1950s AJS 350cc single.

ABOVE: Aprilia cashed in on their World Championship success of 1994 by producing a street version of the race-winning bike. It was tagged the RS250.

1995 APRILIA RS250

Following Aprilia's World Championship triumph in 1994 it was perhaps inevitable that the company would introduce a street bike of the same capacity, clearly inspired by racing technology. The RS250 is that machine. It is intended to be a lighter motorcycle than some of the larger capacity supersport bikes but designed to handle equally as well. The machine is built around an aluminum and magnesium alloy frame of an inclined double beam type into which the V-twin engine, upside-down forks and alloy swingarm are all fitted. The motorcycle, like many sports bikes, is equipped with a full fairing that enhances the sporting appearance of the motorcycle.

SPECIFICATION
Country of origin: ITALY
Capacity: 249cc (15.18cu. in.)
Engine cycle: 2-stroke
Number of cylinders: 2
Top speed: 131mph (210kph)
Power: 65bhp @ 11,500rpm
Transmission: 6-speed
Frame: Twin spar alloy

ARIEL

This concern, which started in 1898, produced De Dion-engined three-wheelers in Birmingham, England. In 1902 production of single-cylinder White & Poppe-engined motorcycles commenced. By the time of the outbreak of World War I, the company had a range of 498cc (30.37cu. in.) side-valve singles and 998cc (60.87cu. in.) inlet-over-

BELOW: The 1926 497cc (30.30cu. in.) overhead-valve machine seen here was one of the models typical of both Ariel and the era. Overhead valves were becoming widely used.

exhaust V-twins. Expansion of the range continued into the 1920s and an individual who was to become famous within the British motorcycle industry joined the company: Edward Turner. He joined as a technician in 1927 to work under the direction of Chief Designer Val Page.

In 1931 Ariel introduced what is arguably their most famous motorcycle, the Square Four. This machine was powered by an air-cooled, four-cylinder engine which featured four vertical cylinders arranged in a square configuration. It was designed by Edward Turner and displaced 498cc (30.37cu. in.). Later this capacity was increased to 596cc (36.35cu. in.) and then again up to 996cc (60.75cu. in.). The overhead-valve gear was operated by a chain-driven camshaft. Also in Ariel's prewar range were the Red Hunter models. During World War II Ariel produced a 347cc (21.16cu. in.) overhead-valve model, tagged the WNG, for the British Army. The postwar range consisted of Red Hunter singles of 347cc (21.16cu. in.) and 497cc (30.31cu. in.) displacement, 498cc (30.37cu. in.) overhead-valve twins, a 598cc (36.47cu. in.) single and the 997cc (60.81cu. in.) Square Four.

Ariel was a casualty of the decline of the British motorcycle industry in that it became closely linked with BSA. In the 1960s a 49cc (2.98cu. in.) model named the Pixie was introduced, which was not a huge success. Two other models were more successful: the Ariel and the Ariel Arrow. These were 250cc (15.25cu. in.) vertical twin two-strokes and later there was a 197cc (12.01cu. in.) version. Ariel's three-wheeler moped, the Ariel Three, was innovative but cannot be described as a success.

1931 ARIEL SQUARE FOUR

The Ariel Square Four – which very soon picked up the nickname 'The Squariel' – was a great sensation when it was unveiled in 1931. The design of the bike, which had been accomplished by Edward Turner, featured a four vertical cylinder engine with the bores in pairs. When originally released the engine displaced 498cc (30.37cu. in.) and had a chain-driven overhead camshaft. The model stayed in production although it was almost continually refined until 1960. Its displacement was sequentially enlarged,

BELOW: The Square Four (nicknamed 'The Squariel') was probably Ariel's most famous and successful motorcycle and in various forms was manufactured for 30 years. This is a 1937 model.

first to 600cc (36.60cu. in.) and then to 1000cc (61cu. in.). There were changes to the cylinder heads and to the materials from which the engine was cast. The larger variant was the one that was reintroduced after the war. As well as refinements to the engine cycle, parts too were enhanced during the production run; along the way 'The Squariel' acquired both rear and front suspension and telescopic forks.

SPECIFICATION
Country of origin: GREAT BRITAIN
Capacity: 498cc (30.37cu. in.)
Engine cycle: 4-stroke
Number of cylinders: 4
Top speed: 95mph (152kph) (Estimate)
Power: n/a
Transmission: 4-speed
Frame: Steel duplex

BELOW: The Ariel Square Four was redesigned in 1953 and given four separate exhaust pipes to differentiate it from the earlier models. This feature lasted beyond 1955 when this example was manufactured.

1932 ARIEL RED HUNTER MH32

The Red Hunter was the name first given to a sporting version of the 500cc (30.50cu. in.) overhead-valve single, although the basics of the Ariel singles went back to 1927 after Val Page joined the company as a designer. He relocated the magneto behind the engine and had it chain-driven from the camshaft. This feature endured for almost 30 years. The Red Hunter, which was officially introduced in 1932, was a sportier version of the Ariel model VG – a four-valve, vertical-cylindered machine. In deference to its name the tank panels and rim centers were painted red.

SPECIFICATION
Country of origin: GREAT BRITAIN
Capacity: 346cc (21.10cu. in.)
Engine cycle: 4-stroke
Number of cylinders: 1
Top speed: 75mph (120kph) (Estimate)
Power: n/a
Transmission: 4-speed
Frame: Steel duplex cradle

LEFT: The Red Hunter was a name that Ariel gave to several machines including this 1952 single-cylinder machine. Trials and scrambles versions were also made.

1963 ARIEL LEADER

The Ariel Leader and Ariel Arrow were machines built in response to the enormous popularity enjoyed by scooters in the 1960s. They were an attempt to combine a motorcycle with the convenience aspects of the scooter, namely the better weather protection which was afforded by a windshield and legshields, as well as keeping the mechanical parts hidden away behind bodywork. The Leader had much more bodywork than the Arrow but both motorcycles were noticeable in that they featured forward-looking styling. The Leader found favor with police departments due to of its panniers and their large carrying capacity.

SPECIFICATION
Country of origin: GREAT BRITAIN
Capacity: 247cc (15.06cu. in.)
Engine cycle: 4-stroke
Number of cylinders: 1
Top speed: 70mph (112kph)
Power: 17.5bhp @ 6750rpm
Transmission: 4-speed
Frame: Pressed steel beam

ABOVE: The Ariel Leader, such as this 1963 model, was a 'cross' between a motorcycle and a scooter; a motorcycle with bodywork.

ARMSTRONG

In 1980 a company which was already a supplier of components to the motor trade made a move into motorcycle manufacture. They did this by purchasing the assets of two other small manufacturers, namely CCM and Cotton. CCM were noted for being off-road specialists and Armstrong built a motorcycle with an Austrian Rotax engine aimed at military contracts. The resultant bike was tagged the MT500, a reference to its almost 500cc displacement (485cc/29.58cu. in.). Later on, Armstrong stopped producing these machines and Harley–Davidson then acquired the manufacturing rights.

B

BAJAJ AUTO

This is the largest scooter and three-wheeler manufacturer in India. In the years prior to 1971 Bajaj assembled scooters under license from Piaggio of Italy using mostly imported components. This arrangement was refined to the extent that Bajaj themselves went on to license the production of their products in Taiwan and Indonesia.

1971 BAJAJ CHETAK

The 1971 model Chetak was the first wholly Indian-built scooter from this company. Scooters are enormously popular in India as a result of their relative cheapness and the huge unsatisfied demand for personal transportation.

SPECIFICATION
Country of origin: INDIA
Capacity: 145.45cc (8.87cu. in.)
Engine cycle: 2-stroke
Number of cylinders: 1
Top speed: 56mph (90kph)
Power: 6.3bhp @ 5200rpm
Transmission: 4-speed
Frame: Steel tubular

BATAVUS

This is a Dutch company that was formed in 1904 and today is best known for small capacity mopeds. Andriess Gaastra founded the company originally to sell clocks and sewing machines. Later on he sold imported German bicycles. Production of motorcycles was started in the 1930s and included a 125cc (7.62cu. in.) model. Batavus is now the leading Dutch manufacturer of mopeds and during expansion has acquired other Dutch companies such as Magneet in 1959 and Phoeniox–Fongers–Germaan in 1970.

BELOW: Batavus have become famous for producing a wide range of small capacity machines intended as cheap and basic commuter transport. This HS50 was a 1974 moped model.

ABOVE: Another type of machine manufactured by Batavus is the step-through moped such as this 1974 Amigo model.

1981 BATAVUS MONDIAL

This was a machine completely typical of Batavus products: a small capacity moped intended purely as local transportation.

SPECIFICATION
Country of origin: HOLLAND
Capacity: 49cc (2.99cu. in.)
Engine cycle: 2-stroke
Number of cylinders: 1
Top speed: 29mph (48kph)
Power: 2.4bhp @ 5000rpm
Transmission: Variable gearing
Frame: Tubular steel

BENELLI

Benelli is the surname of a number of brothers who founded a company to design and build motorcycles in Pesaro, Italy. They started out by manufacturing small capacity machines in 1911 and produced numerous models and designs up until World War II. Their bikes were very successful in the small capacity racing classes. In 1939 Ted Mellors rode a 250cc (15.25cu. in.) Benelli to victory in the Isle of Man Lightweight TT. The factory was destroyed during World War II but it was eventually reopened in 1949

when the production of small capacity machines continued.

A change of policy occurred in the 1970s when Alessandro de Tomaso, who already ran the Moto Guzzi factory, took over Benelli. This resulted in close co-operation between the two companies and later with Motobecane from France. The change of management saw the production of larger machines including the 748cc (45.62cu. in.) six-cylinder Benelli Sei in 1974.

ABOVE: In 1934 the Benelli brothers produced machines such as this overhead-camshaft 175cc (10.67cu. in.) motorcycle.

1974 BENELLI SEI

As in-line four-cylinder engines fitted both along and across the frame of the motorcycle were proven as both functional and practical, it was perhaps inevitable that a designer somewhere would produce an in-line six-cylinder machine. The first such motorcycle came from the automotive industrialist Alessandro de Tomaso and the Benelli company. This extravagant machine was launched with a displacement of 750cc but by the end of the decade the engine had increased to 906cc (55.26cu. in.). The in-line six-cylinder concept was further developed by two of the Japanese manufacturers, namely Honda and Kawasaki.

BELOW: The Benelli Sei of 1974 was unusual in that it had six cylinders in an across-the-frame configuration. Honda manufactured a similar but larger capacity six in the same decade.

SPECIFICATION
Country of origin: ITALY
Capacity: 748cc (45.62cu. in.)
Engine cycle: 4-stroke
Number of cylinders: 6
Top speed: 118mph (190kph)
Power: 71bhp @ 8900rpm
Transmission: 5-speed
Frame: Duplex cradle

1979 BENELLI 254

Benelli had long been interested in multi-cylinder machinery and developed an in-line four-cylinder machine that displaced 250cc (15.25cu. in.) for racing as early as 1940. This Benelli 250cc machine (known as the 254) was one of a series of four-cylinder machines of varying capacities including ones of just over 600cc (36.60cu. in.) and one of 350cc (21.35cu. in.).

BELOW: Transverse four-cylinder engines were more generally accepted for motorcycles and in 1979 Benelli produced the 254. They also produced the 2502C (shown), a two-stroke, vertical twin of 231.3cc displacement.

SPECIFICATION
Country of origin: ITALY
Capacity: 231.1cc (14.09cu. in.)
Engine cycle: 4-stroke
Number of cylinders: 4
Top speed: 90mph (150kph)
Power: 28bhp @ 10,500rpm
Transmission: 5-speed
Frame: Open tubular

BETAMOTOR

This Italian company began in 1904. It started by producing pedal cycles and did

BELOW: The 1985 TR 32 Beta was a motorcycle specifically designed for trials competition.

not move into the production of mopeds and motorcycles until the 1950s. Betamotor employed a designer called Giuseppe Bianchi and produced a range of machines of varying capacities under 250cc (15.25cu. in.). Betamotor are noted for producing both two- and four-strokes, particularly for use in off-road competition.

1986 BETA KR250

Typical of 1980s motocross machinery is the KR250 and its smaller variant the KR125. It is a fast two-stroke fitted to a box section frame that has good ground clearance, long travel leading axle forks and monoshock rear suspension.

SPECIFICATION
Country of origin: ITALY
Capacity: 239cc (14.57cu. in.)
Engine cycle: 2-stroke
Number of cylinders: 1
Top speed: 85mph (140kph)
Transmission: 6-speed
Frame: Box section cradle

BIANCHI

The first motorized machine made by Edoardo Bianchi was a powered bicycle. This was in 1897 and by 1900 he had also constructed a motor car. His enthusiasm for motorcycles, however, remained undiminished and he offered machines with innovative features such as leading link forks (1905). By 1910 he was building a 498cc (30.37cu. in.) single-cylinder machine. This bike was considered reliable and well engineered, and it was the motorcycle which secured Bianchi's position as a successful and respected manufacturer.

In 1916 the company produced a 650cc (39.65cu. in.) V-twin. Through the 1920s Bianchi produced a range of singles of V-twins of various capacities including a double overhead camshaft, 348cc (21.22cu. in.) machine, designed by the Chief Engineer, Albino Baldi, and intended for racing. This motorcycle became one of Italy's most successful racers of that era.

World War II intervened in the success story but the company returned to motorcycle production after 1945 with 123cc (7.50cu. in.) and 248cc (15.12cu. in.) single-cylinder, overhead camshaft machines. Throughout the 1950s the company developed a pair of overhead camshaft, vertical-twins of 248cc (15.12cu. in.) and 348cc (21.22cu. in.) displacement. The company ceased motorcycle production in 1967.

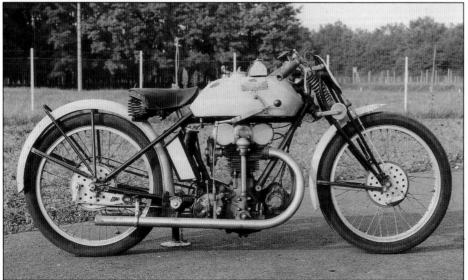

LEFT: Bianchi's 350cc (21.35cu. in.) double overhead camshaft, single-cylinder machine, designed by Albino Baldi, became one of Italy's most successful racing motorcycles.

BIMOTA

Bimota is a small Italian company which builds bespoke sports bikes based around other manufacturers' engines. This might be relatively uncommon now but once was accepted practice within the motorcycle industry. The company's machines are built in small numbers and are famed for being innovative. An example of such innovative engineering is that of the 1982 Bimota Tesi prototype. It was displayed at the Milan Show and comprised a Honda V4 engine in a carbonfiber frame, and in place of a telescopic front fork assembly it featured an alloy swingarm-style wishbone with the steering controlled by hydraulics. The bike had been still further refined by the time it went into production in 1990.

1991 BIMOTA TESI 1D

Bimota moved from Japanese engines to those from another Italian company, Ducati. For the Tesi 1D they chose the state-of-the-art 90° V-twin. This is a double overhead camshaft, liquid-cooled engine with fuel injection. This engine was then

BELOW: The 1991 Bimota Tesi 1D is powered by a Ducati V-twin, liquid-cooled and fuel-injected engine, which is partially concealed by the bodywork. It features alloy swinging arm suspension, both front and rear, and hydraulic steering. An exotic motorcycle.

installed in an innovative chassis that features alloy swinging arms both front and rear. The motorcycle has a large area of bodywork that encloses the engine and forms an integral seat and tank unit.

SPECIFICATION
Country of origin: ITALY
Capacity: 906cc (55.26cu. in.)
Engine cycle: 4-stroke
Number of cylinders: 2
Top speed: 155mph (250kph)
Power: 113bhp @ 8500rpm
Transmission: 6-speed
Frame: Aluminum alloy diamond

BMW

BMW – Bayerische Motoren Werke – was incorporated in 1917, starting as an aircraft engine manufacturer; indeed, their logo to this day is a representation of a spinning propeller. In 1921, the company produced a motorcycle engine for other companies who built motorcycles. This engine was of the configuration generally referred to as a

BELOW: The BMW R1100RS SE is a sports/touring motorcycle powered by an overhead-valve folat twin engine. It produces 90bhp and is capable of 135mph (217kph).

ABOVE: A BMW R11, this was a side-valve machine produced alongside the overhead-valve R16 in the Depression years. Both had pressed steel frames and flat-twin engines.

flat-twin, i.e. it was a horizontally-opposed, twin-cylinder engine. In 1923, prohibited from the manufacture of aeroengines by the Treaty of Versailles, BMW produced a complete motorcycle designed by an aircraft engineer named Max Friz. This motorcycle, which was known as the R3, also featured a side-valve, flat-twin engine of 500cc (30.50cu. in.) and flat-twins became the basis for the majority of BMW motorcycles until the introduction of four-cylinder machines around 50 years later.

In 1929 BMW claimed their first record: Ernest Henne rode a partially-faired, supercharged overhead 750cc (45.75cu. in.) twin at an average speed of 135mph (217kph) for the 'flying mile'. Subsequently he took a further 75 records including the fastest prewar ride of 174mph (280kph), a record which would stand for 14 years. Before World War II supercharging was permitted for road racing and in 1939 Georg Meier won the Senior TT on the Isle of Man on a supercharged BMW.

In the years following the war BMW concentrated more on sidecar racing and it is generally accepted that BMW's flat-twin engine is more suited to sidecar use because engine width is no disadvantage in a three-wheeler, the cylinders are ideally located for cooling, and the shaft drive is suited to this application. As a result numerous race victories have gone to BMW-powered outfits.

1928 BMW R52

The 'R' in BMW model designations is for 'Rad' which is the German for the word 'Cycle'. BMW started producing flat-twin

BELOW: The BMW R12 of 1935 was an R11 fitted with telescopic forks and four-speed transmission. From 1938 until 1942 it was manufactured exclusively for the Wehrmacht.

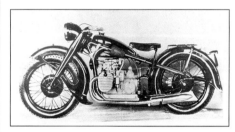

motorcycles in 1923 and for the remainder of that decade the company would offer both side-valve and overhead-valve flat-twins in models with designations that started with the letter 'R'. The R32 was the first side-valve and the R37 became an overhead-valve machine. By 1928 the same engine types were the models R52 and R57 respectively.

SPECIFICATION
Country of origin: GERMANY
Capacity: 494cc (30.13cu. in.)
Engine cycle: 4-stroke
Number of cylinders: 2
Top speed: 70mph (112kph)
Power: 12 PS @ 4000rpm
Transmission: 3-speed
Frame: Twin tube

1941 BMW R75

BMW was the largest manufacturer to supply motorcycles to the Wehrmacht after the military, in 1938, issued a specification for a sidecar outfit intended for cross-country use. The company competed with Zundapp and spent two years working on the development of the machine which became the R75. It was designed to be capable of running for long periods at low speeds to allow for off-road and convoy use. Production started in June 1941 and almost 18,000 machines were produced for the German Army by 1944.

SPECIFICATION
Country of origin: GERMANY
Capacity: 745cc (45.44cu. in.)
Engine cycle: 4-stroke
Number of cylinders: 2
Top speed: 58mph (93kph)
Power: 26 PS @ 4000rpm
Transmission: 4-speed plus reverse
Frame: Steel tube

BELOW: The R52 of 1928 was an upgraded R42. Still powered by a side-valve, flat-twin engine, it featured revised cylinder heads, electric lighting and improved gearshift and brakes.

ABOVE: The BMW R100GS is a dual purpose on-/off-road tourer. It features an overhead-valve, flat-twin engine that produces 70bhp and is capable of 115mph (185kph). The bike is shaft-driven with five-speed transmission.

1952 BMW R68

BMW's future was in doubt after the end of World War II. Their Munich plant was destroyed and their remaining plant found itself inside the Russian Occupied Zone. However, by dismantlng a prewar machine to take measurements – all drawings having been lost – the company once again was able to start producing motorcycles. The first new postwar twin was the 1950-51 R51/2 which was a 500cc (30.50cu. in.) overhead-valve machine. In October of 1951 a new model was displayed at the Frankfurt Motorcycle Show. This was the R68, the new flagship of the BMW range.

SPECIFICATION
Country of origin: GERMANY
Capacity: 594cc (36.23cu. in.)
Engine cycle: 4-stroke
Number of cylinders: 2
Top speed: 100mph (161 kph)
Power: 35 PS @ 7000rpm
Transmission: 4-speed
Frame: Twin tube

ABOVE: The BMW R68 of 1952 was not BMW's first postwar twin but the first sporting one.

1980 BMW R80GS

BMW continued manufacturing roadgoing flat-twin motorcycles into the 1980s but clearly kept an eye on the off-road scene in view of their much earlier successes in events such as the International Six Days' Trial (ISDT). In 1980 the company introduced the R80GS, which was an off-road model but based on the road machine. It featured a 21in. (530mm) front wheel, mudguards with enhanced clearance, and a single-sided rear swingarm. The revised chassis geometry, the high ground clearance and the lighter weight soon made it a firm favorite amongst BMW owners.

SPECIFICATION
Country of origin: GERMANY
Capacity: 798cc (48.67cu. in.)
Engine cycle: 4-stroke
Number of cylinders: 2
Top speed: 115mph (185kph)
Power: 50 PS @ 6500rpm
Transmission: 5-speed
Frame: Twin tube

ABOVE: The R80GS BMW of 1980 was intended as a long distance tourer and was later available in Paris-Dakar rally versions.

1983 BMW K-SERIES

The first major shift from flat-twin engines for BMW in 60 years came in 1983 when the company unveiled the K-Series. It featured an innovative in-line four-cylinder engine that was positioned longitudinally into the frame retaining the shaft drive to the rear hub. The cylinders were horizontal and supplied with fuel through an injection set-up. The bike featured a single-sided alloy swinging arm, rear suspension and telescopic forks.

SPECIFICATION
Country of origin: GERMANY
Capacity: 987cc (60.20cu. in.)
Engine cycle: 4-stroke
Number of cylinders: 4
Top speed: 137mph (220kph)
Power: 90 PS @ 8000rpm
Transmission: 5-speed
Frame: Tubular space

BÖHMERLAND

These machines were designed by Albin Liebisch in an altogether unconventional way. Unlike most other motorcycles they were very long and some models had seating for three people. The forks were unusual in that, although they were leading link, they featured springs under tension. The wheels were novel in that, for the first

BELOW: The Böhmerland was a Czech-manufactured design of motorcycle that was wholly unconventional in appearance.

time, they were cast from a light alloy – something which has become more and more commonplace.

1927 BÖHMERLAND

All machines which were made between 1925 and 1939 were fitted with a single-cylinder, overhead-valve engine with open pushrods and valve gear. The fuel tank was not mounted over the engine in the conventional fashion but two were provided and located either side of the rear wheel.

SPECIFICATION
Country of origin: CZECHOSLOVAKIA
Capacity: 598cc (36.47cu. in.)
Engine cycle: 4-stroke
Number of cylinders: 1
Top speed: 59mph (95kph)
Power: 16bhp @ 3000rpm
Transmission: n/a
Frame: Tubular steel

BRIDGESTONE

This Japanese company was part of the tire company of the same name and started making motorcycles in the early 1950s. Bridgestone began with two-stroke mopeds but soon moved on to slightly bigger singles and twins. They made a 98cc (5.97cu. in.) single and a range of 173cc (10.55cu. in.), 247cc (15.06cu. in.), and 348cc (21.22cu. in.) vertical twins, all of which were two-stroke machines built with engines of Bridgestone's own manufacture. Their main market was the USA due to the fact that they supplied tires to their domestic com-

ABOVE: The 1967 Bridgestone 350 was one of the two-strokes which had its engine built by the company prior to its concentrating on tires.

petitors, the so-called 'Japanese Big Four'. It was concerns over loss of such lucrative tire business that caused Bridgestone to withdraw from motorcycle manufacture in the early 1970s.

BROCKHOUSE

Brockhouse were an engineering company who manufactured motorcycles after World War II in Southport, England. They produced a version of the folding 98cc (5.97cu. in.) Corgi mini-scooter that had been used by British airborne troops during the war. However, Brockhouse eventually became closely involved with Indian Motorcycles from Massachusetts, USA, and built the 248cc (15.12cu. in.) side-valve, vertical single Indian Brave.

BELOW: The Brockhouse Corgi, such as this 1949 model, was in many ways the forerunner of machines such as the Honda Minibike. The Corgi was originally intended for military use.

BROUGH

William E. Brough designed and built motorcycles from 1908 in both single and V-twin engine designs. With the willing collaboration of Granville Bradshaw, he shifted his attention to the use of flat-twin engines – first Bradshaw's own 496cc (30.25cu. in.) ABC engine and later similar configuration engines of 496cc (30.25cu. in.), 692cc (42.21cu. in.) and 810cc (49.41cu. in.) displacement. Production of motorcycles continued until 1926, although Brough's son, George, didn't favor flat-twins and had left the family business to start his own motorcycle company in 1921.

BROUGH–SUPERIOR

Brough–Superior was the name given by George Brough to his company, set up in 1921 in Nottingham, England. His machines were built to the highest standards almost regardless of costs. The motorcycles built by this company were in the main based around V-twin engines supplied by other British manufacturers such as Matchless, JAP and MAG. Possibly George Brough's most famous customer was T .E. Lawrence (of Arabia) who was killed on a Brough–Superior in Dorset, England, in the 1930s.

RIGHT: There is nothing that can be really described as the standard Brough–Superior as each was built to some degree with the buyer in mind. This model was made in 1925.

BELOW: The 1936 Brough–Superior SS80 seen here was powered by a Matchless engine although it was marked as a Brough. The engine displaced 982cc (59.20cu. in.).

Production continued until 1940 when World War II prevented further manufacture. The company did not start up again after the war had ended because Brough did not think any of the engines which were available at that time would be suitable for use in his motorcycles.

1930 BROUGH–SUPERIOR SS100

George Brough knew that he could charge high prices for quality motorcycles but that he had to keep his customers happy through prompt dispatch of spare parts. The SS100 was one of his machines which was assembled to the highest standards. It used a V-twin engine from JAP and featured both front and rear suspension.

ABOVE: The Brough–Superior SS100, such as the 1939 example shown, was always the top of the range model from the company hailed as 'the Rolls–Royce of motorcycles'. War halted production and Brough went on to make crankshafts for Rolls–Royce aeroengines.

SPECIFICATION
Country of origin: GREAT BRITAIN
Capacity: 995cc (60.69cu. in.)
Engine cycle: 4-stroke
Number of cylinders: 2
Top speed: 100mph (161kph)
Transmission: 3-speed
Frame: Spring

BSA

BSA stands for Birmingham Small Arms who, a long time before they manufactured motorcycles, mass-produced small arms in Birmingham, England. The company was founded by a group of craftsman gunsmiths who saw the benefits of mass-production as military requirements increased dramatically and weaponry was needed in quantities never before envisaged. BSA produced its first motorcycle in 1910. This was a single-cylinder machine with belt drive. From this humble beginning, BSA went on to greater things. By 1920 there were three models in the range including a V-twin of 770cc (46.97cu. in.) displacement, which was seen as ideal for pulling sidecars. Harry Perry rode a 350cc (21.35cu. in.) 1924 model to the top of Mount Snowdon in Wales as a publicity stunt. After an embarrassment in the 1921 TT, BSA shied away from racing and preferred to concentrate on trials and reliability tests. The company concentrated on producing bikes that ordinary people would buy to ride to work and it was this philosophy which helped them to survive the Depression of the 1930s.

In addition to supplying rifles to the British Army during World War II, BSA produced over 120,000 of the M20, a side-valve 500cc (30.50cu. in.) motorcycle for

ABOVE: BSA produced almost 126,000 496cc (30.25cu. in.) single-cylinder motorcycles, such as this 1942 M20, for the British war effort. The side-valves acquired a reputation for being reliable and were used in all theaters of operations, including deserts.

ABOVE: The prewar-designed M20 stayed in production after World War II – this is a 1946 model. It was superseded by the 591cc (36.05cu. in.) side-valve M21 which retained girder forks, although telescopic ones were fitted in other BSA machines.

the Allied armies. In order to expand production during the war, BSA acquired Sunbeam, New Hudson and Ariel. In the postwar years the company first unveiled bikes such as the A7, which was a semi-unit parallel-twin. By 1948 the company's range was comprehensive and ranged from utility sidecar machines (the M-series) to sporting machines, and from 125cc (7.62cu. in.) motorcycles to 650cc (39.65cu. in.) models. Through the 1950s it seemed that BSA could do no wrong. Their bikes were popular and sold well. Models like the Gold Star proved successful both in the Isle of Man TTs and in terms of sales. At the end of the decade labor relations saw the beginning of the end for the company which had claimed to manufacture 'the world's most popular motorcycles'. By 1971 things were precarious and, after some difficulties with the sales of shares, the company closed down.

1928 BSA SLOPER

The BSA Sloper, with its inclined engine, was a tuned version of an earlier touring model, and, to ensure it was recognized as a new sporting model, it carried a red star emblem. This was the beginning of a line of models named as various 'stars', including both Empire Stars and Gold Stars.

SPECIFICATION
Country of origin: GREAT BRITAIN
Capacity: 493cc (30.07cu. in.)
Engine cycle: 4-stroke
Number of cylinders: 1
Top speed: 75mph (120kph) (Estimate)
Power: n/a
Transmission: 3-speed
Frame: Tubular steel

LEFT: The 1928 BSA Sloper of 493cc (30.07cu. in.) displacement. It was so described because of the forward sloping cylinder. In 1928 lights such as appear on this model were sold as an extra-cost option.

1946 BSA D1 BANTAM

This model had its roots in the prewar DKW two-stroke, the design of which came to BSA after the defeat of Nazi Germany. The BSA 123cc (7.50cu. in.) version of the unit construction two-stroke was launched in 1946 intended as cheap commuter transport. Subsequently there was a 175cc (10.67cu. in.) version and a number of other refinements in a production run that lasted through the 1960s.

BELOW: The design of the BSA Bantam, a small capacity two-stroke, was based on the German DKW two-stroke, as was the Harley Hummer.

SPECIFICATION
Country of origin: GREAT BRITAIN
Capacity: 123cc (7.50cu. in.)
Engine cycle: 2-stroke
Number of cylinders: 1
Top speed: 54mph (86kph)
Power: 5bhp @ 5000rpm
Transmission: 3-speed
Frame: Tubular steel

1956 BSA GOLD STAR

The origins of the Gold Star model name are interesting and are linked with the historic British race circuit, Brooklands. A gold star lapel badge was awarded to riders who

ABOVE: In the mid-1950s the BSA Gold Stars were built in road, scrambles, road-racing and clubman's forms. This variety was reduced to only two versions by 1957. The DBD34 shown was the final form of the 499cc (30.43cu. in.) Gold Star, seen here in Clubman's trim.

completed a lap of the famous circuit at over 100mph (161kph). Walter Handley did this for BSA on a 500cc single in 1937. For the 1938 season the bike was renamed the Gold Star. The name reappeared in the years after World War II on another 500cc single, a performance all-alloy engine. The postwar Gold Star became a firm favorite with both racers and rockers.

SPECIFICATION
Country of origin: GREAT BRITAIN
Capacity: 499cc (30.43cu. in.)
Engine cycle: 4-stroke
Number of cylinders: 1
Top speed: 110mph (178kph)
Power: 40bhp @ 7000rpm
Transmission: 4-speed
Frame: Tubular steel

1968 BSA ROCKET 3

The Rocket 3 was an acknowledgement by BSA that the superbike era was coming. It was the culmination of attempts by the linked Triumph and BSA companies to produce a large capacity and fast motorcycle. The engine is essentially derived from a Triumph Daytona engine with an extra cylinder grafted on. Triumph produced the Trident and BSA the Rocket 3 because both

is famed for building trials and motocross machines which are popular in both Europe and North America. The company also produced fast, if diminutive, road-racing bikes which were ridden with considerable success by Angelo Nieto.

1980 BULTACO SHERPA T74

This bike was introduced in the early 1960s and immediately made its presence felt in trials competition by rendering much of its competition obsolete. This was because of its delivery of torque at low revs and also due to its hill-climbing ability. Bultaco kept updated Sherpa models in production for over a decade.

SPECIFICATION
Country of origin: SPAIN
Capacity: 74.79cc (4.56cu. in.)
Engine cycle: 2-stroke
Number of cylinders: 1
Top speed: n/a
Power: 6.6bhp @ 7000rpm
Transmission: 6-speed
Frame: Tubular

ABOVE: The BSA Rocket 3 was a three-cylinder bike made by both Triumph and BSA, albeit with minor differences because the two companies had different US importers. To add to the confusion the Craig Vetter-styled X-75 Triumph Hurricane used a BSA Rocket 3 engine.

companies had different US importers. Both the BSA and Triumph variants performed well at speed including a 1971 1-2-3 win in the Daytona 200. Dick Mann and Don Emde took first and third places respectively on BSA machines, with Gene Romero placed second on a Triumph.

SPECIFICATION
Country of origin: GREAT BRITAIN
Capacity: 740cc (45.14cu. in.)
Engine cycle: 4-stroke
Number of cylinders: 3
Top speed: 122mph (196kph)
Power: 60bhp @ 7250rpm
Transmission: 4-speed
Frame: Duplex

BUELL

This American company was set up in the early years of the 1980s by Eric Buell, who is a former employee of Harley–Davidson.

RIGHT: The Buell is a sports bike that uses a Harley–Davidson Sportster engine. The machine was designed and manufactured by Eric Buell, a former Harley–Davidson employee. His innovative approach attracted his former employers who now offer the machines through their own dealerships.

His intention was to use Harley–Davidson engines and fit them to high performance handling frames. The bikes were built in limited numbers and came to be regarded as enthusiasts' machines. In the early 1990s Harley–Davidson took an interest in the sports bike manufacturer and offered the machines through its own dealerships.

BULTACO

This Spanish company was founded in 1958 by Francisco Bulto, a designer and former Montesa employee. He started working from a farmhouse but soon moved to a factory which is based in Barcelona, Spain, and

BELOW: Bultaco, a Spanish company, introduced the Sherpa into trials competition in the 1960s. It was a great success and the machine was refined for several years. This is a 1973 model.

C

CAGIVA

Cagiva Motor – run by brothers Claudio and Gianfranco Castiglioni – took over Varese, the Italian-based Aermacchi, from Harley–Davidson in 1978 and began by producing variants of the existing models. The main changes were detail upgrades and also a new tank badge, as well as improved electrics and cast wheels. It was enough to make the newcomer's bikes the best-selling in Italy in the 125cc (7.62cu. in.) class for several years. Times moved on, however, and Cagiva successfully sought to produce more modern and larger capacity motorcycles. In this process of expansion Cagiva first acquired Ducati in 1985 and then Husqvarna in 1986. They went on to take over Moto Morini in 1987 and CZ in 1993.

1980 CAGIVA SST350

The first bikes from the Cagiva operation were essentially re-badged Aermacchi/ Harley–Davidsons such as the SST350. There was also an SX model which was a dual purpose 'street scrambler'.

SPECIFICATION
Country of origin: ITALY
Capacity: 341.8cc (20.84cu. in.)
Engine cycle: 2-stroke
Number of cylinders: 1
Top speed: 84mph (136kph)
Power: 32bhp @ 6500rpm
Transmission: 5-speed
Frame: Tubular steel

1995 CAGIVA EXPLORER E900

As well as acquiring other manufacturers, Cagiva have developed a range of on- and off-road motorcycles which are marketed under the Cagiva name. The company manufactures two models of the Paris-Dakar style of off-road machine, known as the Explorer. There are 750cc (45.75cu. in.) and 904cc (55.14cu. in.) versions which are very similar in specification, with the exceptions of the engines, transmissions and some other details such as the diameter of brake discs.

SPECIFICATION
Country of origin: ITALY
Capacity: 904cc (55.14cu. in.)
Engine cycle: 4-stroke
Number of cylinders: 2
Top speed: n/a
Power: 68bhp @ 6500rpm
Transmission: 5-speed
Frame: Box section

ABOVE: The Cagiva Elephant was a 1990s bike, styled on the Paris-Dakar type of motorcycle. It was available in sponsor Lucky Strike's colors and the more conventional paint scheme.

CALTHORPE

George Hands founded the Calthorpe concern in 1911 in Birmingham, England, and started by building 211cc (12.87cu. in.) two-stroke machines and then other models with Precision and JAP engines. During the early 1920s a variety of Blackburne, JAP and Villiers engines were used in Calthorpes but from 1925 onward the company manufactured 348cc (21.22cu. in.) and 498cc (30.37cu. in.) singles. Similar engines of the company's own manufacture powered the Calthorpe range of motorcycles through the 1930s, although the company went into liquidation in 1938. The receiver sold the

BELOW: The Calthorpe 494 Major was one of only two models from the company in 1934.

company to Bruce Douglas, nephew of the founder of Douglas Motorcycles, who moved the operation to Bristol. He commenced production of Matchless 347cc (21.16cu. in.) and 497cc (30.31cu. in.) machines but unfortunately few were finished by the outbreak of World War II. The company did not return to motorcycle manufacture once the war had ended, and its machinery was sold.

1930 IVORY CALTHORPE

Calthorpe are best remembered for one particular model, namely the Ivory Calthorpe, although that name was used on more than one model, including a 248cc (15.12cu. in.) two-stroke, the Ivory Minor. The Ivory Calthorpe stayed in production for several years and the design was gradually upgraded, including improvements such as the fitment of a four-speed transmission in 1931.

SPECIFICATION
Country of origin: GREAT BRITAIN
Capacity: 348cc (21.22cu. in.)
Engine cycle: 4-stroke
Number of cylinders: 1
Top speed: n/a
Power: n/a
Transmission: 3-speed
Frame: Tubular steel

ABOVE: The 1975 Can–Am Bombardier was powered by a two-stroke Rotax engine.

CAN–AM

Can–Am are a division of the Canadian Bombardier Group who are famed for the manufacture of snowmobiles, although the motorcycle factory is actually based in Austria and has been producing bikes since 1973. Originally Can–Am was founded in 1942. However, in 1976 it merged with MLW–Worthington. The resulting entity became known as the Bombardier–MLW Group, with the aim of manufacturing both recreation and transportation products. The Rotax-powered motorcycles have gained a reputation for quality and performance and are used by several NATO armies.

CASAL

This Portuguese company started manufacturing in 1966 and used Zundapp engines. Casal later went on to produce a range of their own two-stroke engines ranging from 49cc (2.98cu. in.) to 248cc (15.12cu. in.).

1980 CASAL K185S

This machine is typical of the models made by Casal. The range included enduro, trail and street bikes and mopeds. The K185S is

BELOW: This K185 Mini Trail bike from 1974 displaces 75cc (4.57cu. in.) and is typical of the two-stroke machines made by Casal.

a diminutive trailbike and a variant, the K188 Enduro, was the same but fitted with Marzocchi forks and Magura brakes.

SPECIFICATION
Country of origin: PORTUGAL
Capacity: 49.9cc (3.04cu. in.)
Engine cycle: 2-stroke
Number of cylinders: 1
Top speed: 56mph (90kph)
Power: 6.5bhp @ 8500rpm
Transmission: 5-speed
Frame: Tubular

CIMATTI

In postwar Italy, Cimatti was first established as a bicycle manufacturer based near Bologna. The company was founded by Marco Cimatti who had been an Olympic cyclist in the 1932 Games. His company began manufacturing mopeds in 1950.

1980 CIMATTI KAIMAN SUPER 6

The Cimatti company use Moto Minarelli engines for a number of their products including their step-through style range of mopeds which also feature automatic clutches. The Super 6 is a trail bike typical of the type, with telescopic forks, swinging arm suspension and high ground clearance.

ABOVE: A Cimatti Kaiman Trial of 1977 that displaces 49cc (2.98cu in.). A 125cc (7.62cu. in.) version was also available.

SPECIFICATION
Country of origin: ITALY
Capacity: 49.6cc (3.04cu. in.)
Engine cycle: 2-stroke
Number of cylinders: 1
Top speed: n/a
Power: n/a
Transmission: 6-speed
Frame: Tubular

CLEVELAND

Cleveland were one of the plethora of short-lived American motorcycle manufacturers. They were in business between 1915 and 1929, during which time they took over Reading Standard in 1922. The motorcycles which were built by Cleveland are acknowledged to have been of a sound design and started with a 269cc (16.40cu. in.) two-stroke. In 1924 the company produced a 347cc (21.16cu. in.) single with an inlet-over-exhaust engine, and then a 746cc (45.50cu. in.) in-line four. This latter machine was further refined for 1928 and increased in capacity to 996cc (60.75cu. in.) but it was unable to compete effectively with the Henderson and Ace Fours. The Wall Street Crash of 1929 forced the company out of business, as it did so many others at that time, despite the high quality of their products.

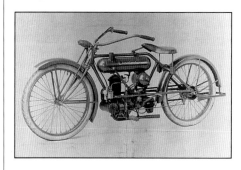

ABOVE: The Cleveland motorcycle of 1919 was typical of many American makers of the time. Few of these concerns survived the Depression.

1923 CLEVELAND MODEL 20

When this machine was introduced it offered a number of original and innovative features including a clutchless gear-change and a worm primary drive. Ignition was accomplished by means of a directly-driven magneto.

SPECIFICATION
Country of origin: USA
Capacity: 269cc (16.40cu. in.)
Engine cycle: 2-stroke
Number of cylinders: 1
Top speed: 40mph (64kph)
Power: 2.5bhp
Transmission: 2-speed
Frame: Rigid steel tube

ABOVE: The Clyno motorcycle of 1914 featured belt drive and clearly demonstrated its bicycle heritage. The proprietary two-stroke engine produced 2.25hp.

CLYNO

Clyno was an English company, based in Wolverhampton, which built both cars and motorcycles. Prior to World War I the latter vehicles had engines supplied by the Stevens brothers who were owners of the Wolverhampton-based AJS factory. After the war was over the company produced 269cc (16.40cu. in.) two-strokes and 925cc (56.42cu. in.) side-valve V-twin models. In 1924 it was decided that, in view of the increasing demand for Clyno cars, production of motorcycles should be discontinued.

COSSACK

Cossack is one of the names given to the motorcycles produced and exported by the Russian state-owned plants that include those at Dniepr, Minsk, Kovrov, Irbit, Izhevsk, Lvov and Riga. Each plant special-

BELOW: A 1975 Cossack 650D. The D suffix denotes Dniepr, a state-owned motorcycle plant in the former USSR. The bike is powered by a four-stroke 650cc (39.65cu. in.) flat-twin.

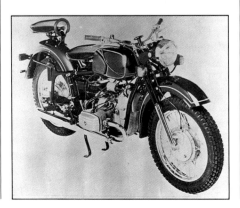

izes in a single model. The latter two factories, for example, produce mopeds; Dniepr build a motorcycle based on the shaft-drive BMW design; Minsk build 125cc (7.62cu. in.) two-strokes based on the DKW design; Kovrov build a 175cc (10.62cu. in.) derivative of the DKW design; Irbit also build a BMW-inspired flat-twin which in some export markets is known as the Ural due to the close proximity of the plant to the Ural mountains; Izhevsk build a single-cylinder two-stroke called Planeta and a twin called the Jupita.

Motorcycle production in the USSR began in 1930 in Leningrad. The flat-twins have their parentage in German motorcycles following the Russian occupation of part of Germany after the collapse of the Third Reich. The motorcycles are considered basic and primitive by many but attract enthusiasts wherever they are marketed in both the USA and in Europe.

1977 DNEIPR MT1O-36

This motorcycle is based around a flat-twin engine configuration and, like several other bikes, has its roots in the motorcycles produced in Nazi Germany before and during World War II. While many of these designs have passed into history, the early type of flat-twin has survived in the USSR. Given the changing political situation and the increasing Western investment in former Eastern Bloc countries, its future after 60 or so years could finally be uncertain.

SPECIFICATION
Country of origin: USSR
Capacity: 649cc (39.58cu. in.)
Engine cycle: 4-stroke
Number of cylinders: 2
Top speed: 81mph (130kph)
Power: 36 bhp @ 5200rpm
Transmission: 4-speed
Frame: Tubular steel

COTTON

Cotton was a company based in Gloucester, England, which began manufacturing motorcycles in 1920. It took its name from the surname of the founder, Francis Willoughby Cotton. The company's company was noted for its triangulated design of frame which endured until 1939. The marque had considerable Isle of Man TT success including Stanley Woods' 1923 Junior TT victory. This triumph went a long way toward establishing the company in the public eye. The Depression, however, brought hard times and, in order to stay in business, Cotton began to offer proprietary engines from companies such as JAP and Blackburne in an extensive range of models. The company closed down in 1939.

ABOVE: A 1930 Model 255A Cotton. This machine displaced 495cc (30.09cu. in.).

COVENTRY–EAGLE

The company was located in Coventry, England, and had its roots in the Victorian era of bicycling. Coventry–Eagle always used proprietary parts but through careful assembly and a high standard of finishing the company survived longer than a number of their competitors. During the 1920s they offered both the luxurious Flying 8 series, which used V-twin JAP engines and sculptured tanks, and utterly utilitarian machines in pressed steel frames.

ABOVE: A 249cc (15.18cu. in.) Coventry–Eagle Pullman model N11 of 1937 with semi-elliptic rear springs.

CROCKER

This is another American marque and another which took its name from that of its founder. In this case he was Al Crocker, a Los Angeles, USA, Indian dealer. He began manufacturing motorcycles in 1933, intent on competing against the British bikes on

ABOVE: Crocker motorcycles earned a reputation for being fast but their machines were produced in very limited numbers in the years before World War II.

the very active speedway racing scene. By 1935 he had moved on to a more typical American engine in the guise of a 45° V-twin featuring overhead valves and limited manufacture began. However, World War II halted production.

CUSHMAN

This American company from Lincoln, Nebraska, USA started producing basic scooters in 1937. These were powered by an industrial engine that displaced 222cc (13.54cu. in.). Cushman produced a number of models, essentially variations on this

BELOW: The Cushman scooter was built as basic transportation but featured some styling touches from bigger American machines.

theme, including some for the US Army in World War II. Production halted in 1965.

CYCLONE

The Cyclone was produced by Joerns Manufacturing from St Paul, Minnesota, USA, for only four years but in that short period became highly acclaimed. The V-twin that was introduced in 1913 had overhead camshafts and displaced 996cc (60.75cu. in.). It was the most technologically advanced motorcycle of its day featuring as it did roller bearing crank and con rods, a bevel-driven camshaft, and forged steel flywheels of a spoked design. The machining was so accurate that shims were not required to meet maximum end float of 0.001ins (0.00254mm) for the cam drive. The performance of a Cyclone was notably fast for the times and the machine almost reached legendary status, especially after the company went out of business at the end of the 1916 race season.

ABOVE: The Cyclone was a fast motorcycle with advanced engineering for its time, including the spring frame seen on this 1915 model.

CZ

CZ stands for Ceska Zbrojovka, the Czech arms factory. It originally belonged to Skoda and started building motorcycles in 1932. The first machines were 73cc (4.45cu. in.) and 98cc (5.97cu. in.) displacement lightweights although production of the 173cc (10.55cu. in.) and 248cc (15.12cu. in.) machines soon followed. These were two-strokes with pressed steel frames. World War II intervened and, in 1945, CZ became nationalized as did Jawa. The two companies subsequently became both technically and commercially connected. In the postwar years the Strakonice, Czechoslovakia, factory produced a variety of motorcycles with capacities ranging from 123cc (7.50cu. in.) to 348cc (21.22cu. in.).

1991 CZ 125

The state-produced motorcycles were only updated and refined sparingly and retained the same basic configuration for many years. Many of the improvements were cosmetic and others were to satisfy legislation in the countries to which CZ products were exported. As well as the 123.7cc (7.54cu. in.) version, there was another model that displaced 175cc (10.67cu. in.).

SPECIFICATION
Country of origin: CZECHOSLOVAKIA
Capacity: 123.7cc (7.54cu. in.)
Engine cycle: 2-stroke
Number of cylinders: 1
Top speed: 52mph (85kph)
Power: 11.5bhp @ 5750rpm
Transmission: 4-speed
Frame: Tubular cradle

BELOW: The 1976 CZ Enduro. Off-road competition was one area where CZ excelled.

ABOVE: The Crocker was a sports machine built by a Californian motorcycle dealer, Al Crocker, in the 1930s. His motorcycles had a reputation for being fast but were built by hand in extremely limited numbers.

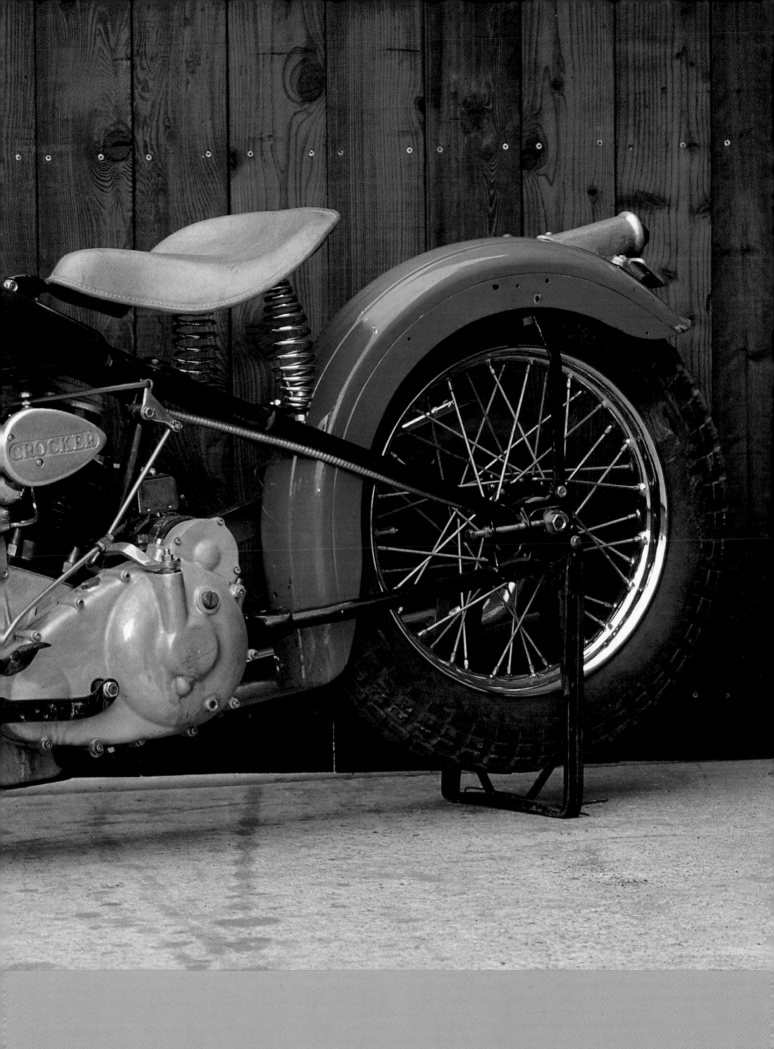

D

DERBI

Simeon Rabasa Singla founded this Spanish company in 1922. It was a family business which manufactured and hired out bicycles. Motorcycle production did not begin until 1950 when a range of lightweights were built, known as SRS after the founder. By 1955 the range included the 125cc (7.62cu. in.), the 250cc (15.25cu. in.) and the 350cc (21.35cu. in.) models and used the name Derbi. This Barcelona-based company has since concentrated on small capacity motor-cycles and mopeds.

BELOW: The Senda of 1995, Derbi's small capacity (50cc/3.05cu. in.) motocross style of machine. It features long travel suspension.

1976 DERBI TRICAMPEONA

In the late 1970s Derbi were producing around 80,000 motorcycles and mopeds annually in their Barcelona factory. The Tricampeona is typical of this production. It features a small capacity two-stroke, single-

ABOVE: The Derbi Tricampeona was so named after Angelo Nieto's three successive World Championships aboard a Derbi machine.

cylinder engine in a completely traditional frame and with traditional parts. Telescopic forks and a rear swinging arm assembly, as well as drum brakes, mean that it is simply cheap and economical transport.

SPECIFICATION
Country of origin: SPAIN
Capacity: 48.767cc (2.97cu. in.)
Engine cycle: 2-stroke
Number of cylinders: 1
Top speed: 25mph (40kph)
Power: 4.5bhp @ 5700rpm
Transmission: 4-speed
Frame: Welded steel

DKW

The DKW concern was founded in Zschopau, Germany, in 1919 by Skafte Rasmussen, an expatriate Dane. The first machines built were simple and reliable two-stroke engines fitted to bicycle-like frames. From such humble beginnings mushroomed a company that, by the 1930s, was the world's largest motorcycle manufacturer. Their extensive racing experi-ence gave DKW the lead in two-stroke technology. World War II intervened and, although DKW manufactured a single, the RT125, it was the beginning of the end. The Zschopau factory site ended up in the Russian Occupied Zone after the end of the war, and the company relocated to the West with some of the original personnel. The Zschopau site became the base for MZ, the renowned two-stroke manufacturer. Production of the RT125 continued after the move although two of the Allied pow-ers – the USA and Great Britain – took the design of the RT125 and it subsequently became the basis of the BSA Bantam and the Harley Hummer. DKW sales declined steadily during the 1950s and the operation became absorbed into another company in the 1960s.

DKW's innovative two-stroke technology brought them fame in the Isle of Man Lightweight (250cc/15.25cu. in.) Races. In 1936 Stanley Woods set a record lap time before being forced to retire, while in 1938 Ewald Kluge achieved victory. They attained their success through the use of super-charging and due to a design referred to as 'split-single'. This had two bores, one in front of the other, in a single casting and with a single common combustion cham-ber. The transfer ports were in the forward bore while the exhausts were in the rear. The pistons were articulated on different length con rods allowing maximum advan-tage in terms of cylinder-filling and valve-

LEFT: The 1935 DKW SB500 featured a two-stroke engine, pressed steel forks and frame, all of which were typical of German machines.

ABOVE: Despite Isle of Man TT success, most DKWs were utilitarian like this 1938 model.

opening. The engine, which was supercharged, breathed through a rotary valve which pushed the power output up further. Germany was not admitted to racing until 1951 by which time further research was in vain because of the introduction of the postwar ban on superchargers.

In 1966 DKW was merged with Express, Victoria and Hercules to become the conglomerate known as Zweirad Union, the biggest manufacturer of two-wheelers in Germany. This, in turn, became a part of the Sachs Group where the name of DKW endured until the 1970s.

1950 DKW RT125

This machine was the postwar reincarnation of the prewar motorcycle which had an almost identical specification and was one of the first models reintroduced after the interruption caused by World War II.

SPECIFICATION
Country of origin: GERMANY
Capacity: 123cc (7.50cu. in.)
Engine cycle: 2-stroke
Number of cylinders: 1
Top speed: 52mph (85kph)
Power: 4.5bhp @ 5000rpm
Transmission: 4-speed
Frame: Tubular steel

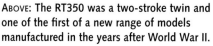

ABOVE: The DKW RT125 was reintroduced in the years after World War II. It also became the basis of BSA and Harley lightweights.

1953 DKW RT350

This was one of DKW's first new postwar designs following their relocation from Zschopau to Ingolstadt. The two-stroke twin was a new design of engine but otherwise it is a completely traditional design of motorcycle.

ABOVE: The RT350 was a two-stroke twin and one of the first of a new range of models manufactured in the years after World War II.

SPECIFICATION
Country of origin: GERMANY
Capacity: 346cc (21.10cu. in.)
Engine cycle: 2-stroke
Number of cylinders: 2
Top speed: n/a
Power: n/a
Transmission: 4-speed
Frame: Tubular steel

1974 DKW W2000 ROTARY

Two-stroke technology was an area of development in which both pre- and postwar DKW factories seemed to excel and an unusual motorcycle appeared from DKW in 1974. This was a rotary-engined machine, the W2000. It was also widely sold as a Hercules, which was another brandname owned by the parent company of DKW. Unfortunately, despite the innovation that went into the design, the motorcycle was not a great success. This was due to a number of factors but mainly to a combination of both its high cost and its less than pleasing appearance.

SPECIFICATION
Country of origin: GERMANY
Capacity: 294cc (17.93cu. in.)
Engine cycle: Wankel rotary
Number of cylinders: 1
Top speed: 90mph (150kph)
Power: 27bhp @ 6500rpm
Transmission: 4-speed
Frame: Tubular steel

LEFT: As part of the Zweirad Union DKW marketed the rotary-engined W2000 in the 1970s. It was also sold under the Hercules brandname. This 1974 model shows that its appearance was not its greatest asset.

DOT

DOT stands for 'Devoid of Trouble' and this British company was founded in 1903 by Harry Reed, a motorcyclist who had been placed in the top three twice – including a 1908 win – at the Isle of Man TT. They stayed in business through the 1920s and in the 1930s diversified into three-wheeler production. The company always used proprietary engines from JAP, Blackburne and, to a lesser extent, Bradshaw. In later years – after their return to motorcycle production in 1949 – they utilized engines from Villiers and, toward the end of the 1960s, Minarelli engines were used.

ABOVE: This 1963 DOT Trials 250 is powered by a two-stroke Villiers engine.

DOUGLAS

Douglas, an English company, were noted for producing motorcycles with flat-twin engines. The first Douglas motorcycle was designed by J. F. Barter. For many years the company produced only flat-twin models although it did manufacture models in a variety of different capacities including 384cc (23.42cu. in.), 498cc (30.37cu. in.) and 596cc (36.35cu. in.). During World War I Douglas received contracts to supply the British Army with motorcycles and, after the war was over, their designs were built under license in Germany by Bosch. In the

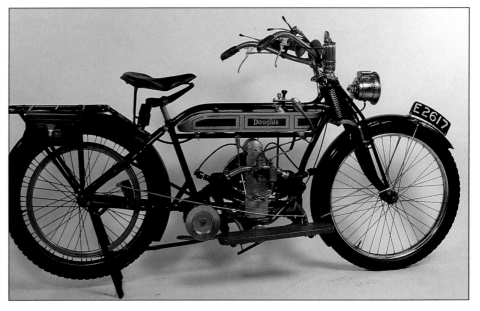

late 1920s the firm built speedway bikes but in the main continued with flat-twins.

The Douglas family relinquished control of the company in 1932 and, after some reorganization, the company attempted to offer a line of less expensive motorcycles which were to be built in larger quantities. The new range of machines stayed with longitudinally-mounted flat-twin engines with the exception of the Endeavour which featured a transversely-mounted flat-twin engine of 498cc (30.37cu. in.).

After World War II Douglas concentrated on the 348cc (21.22cu. in.) flat-twins with transversely-mounted overhead-valve flat-twin engines including the Dragonfly. Also in the postwar years the company, which was part of Westinghouse in Bristol, England, diversified into the distribution of Vespa scooters in the UK. Production of motorcycles stopped in 1956.

BELOW: The 1956 Douglas Dragonfly, powered by a flat-twin engine.

ABOVE: Douglas became noted for the production of flat-twin engines such as the side-valve unit in this 1914 model in which the engine is mounted longitudinally.

1914 DOUGLAS

During World War I many 348cc (21.22cu. in.) side-valve engines of a horizontally-opposed twin configuration were supplied to the British Army. The displacement was achieved through a bore and stroke of 60.8 and 60mm.

SPECIFICATION
Country of origin: GREAT BRITAIN
Capacity: 348cc (21.22cu. in.)
Engine cycle: 4-stroke
Number of cylinders: 2
Top speed: 40mph (64kph)
Power: n/a
Transmission: 2-speed
Frame: Tubular steel

1931 DOUGLAS D31

In 1931 Douglas adopted tartan borders for their gas tank panels but in the main kept their earlier range much as it was. The D31 was a sports/touring machine which was fitted with Douglas's largest capacity side-valve engine. For 1932 their range of models, such as the A32 Terrier, was updated slightly. This featured a 348cc (21.22cu. in.) side-valve flat-twin.

SPECIFICATION
Country of origin: GREAT BRITAIN
Capacity: 596cc (36.35cu. in.)
Engine cycle: 4-stroke
Number of cylinders: 2
Top speed: n/a
Power: n/a
Transmission: 3-speed
Frame: Tubular steel

DUCATI

Ducati was a big Italian industrial company which had to transfer its production to peacetime products at the end of World War II. Ducati realized that there was a need for cheap and basic transportation and, through Aldo Farinelli, started to manufacture the Cucciolo, a 49cc (2.99cu. in.) engine designed to clip on to a normal pedal cycle. Ducati produced large numbers of these machines over a 10-year period which began in 1946. The success of this product helped the company move into small capacity motorcycles and with the arrival of designer Fabio Taglioni progress was steady. Taglioni designed the 100 Gran Sport which pointed the way to the superbike era. However, the company was often split through boardroom wrangles and this had a detrimental effect on production.

Mike Hailwood brought good fortune Ducati's way when, in 1978, aboard a race-prepared 900SS, he won the Isle of Man TT, and this on his dramatic return to the island. His success generated a great deal of good publicity for the Ducati company, which subsequently proceeded to market a Mike Hailwood replica. It was to be Taglioni's last motorcycle design. The Pantah is acknowledged as the motorcycle that saved Ducati from closure and brought them to the attention of Cagiva.

ABOVE: Mike Hailwood's Isle of Man TT race victory aboard a Ducati in 1978 brought much prestige to the Italian manufacturer which went on to offer replicas of the winning bike.

1975 DUCATI GTL 350

Fabio Taglioni, Ducati's Chief Engineer, developed the Desmodromic valve set-up for 125cc (7.62cu. in.) race engines. They would rev to 15,000rpm without harm and achieved great racing success. In the late 1960s Ducati were nationalized and they introduced their roadgoing Desmodromic engines which displaced 250 and 350cc. Despite their cost, these machines were the beginnings of the roadgoing legend. These bikes were soon followed by the 750cc (45.75cu. in.) Ducati 750 GT. It won the 200-miler at Imola.

SPECIFICATION
Country of origin: ITALY
Capacity: 349.6cc (21.32cu. in.)
Engine cycle: 4-stroke
Number of cylinders: 2
Top speed: 105mph (170kph)
Power: 38bhp @ 7000rpm
Transmission: 5-speed
Frame: Tubular steel

1976 DUCATI 860 GTS

The Ducati 860 GTS was introduced in 1976 and was the first bike that indicated the Italian factory could offer a bike to match the Japanese. It was based on the Guigiaro-styled 860 GT but with a different

ABOVE: The 1976 Ducati 860 GTS was a sports bike capable of matching the Japanese. It featured a 90° V-twin engine and was capable of 115mph (185kph).

seat and tank to make it more practical for road use. It is based around Ducati's 90° V-twin engine which allows it to be installed low in the frame, keeping the center of gravity low to enhance handling.

SPECIFICATION
Country of origin: ITALY
Capacity: 864cc (52.70cu. in.)
Engine cycle: 4-stroke
Number of cylinders: 2
Top speed: 115mph (185kph)
Power: 67.7bhp @ 7000rpm
Transmission: 5-speed
Frame: Duplex tubular

1990 DUCATI 900 SUPERSPORT

The 900 Supersport followed hot on the heels of the successful 750 Sport. It used an existing engine – the 904cc (55.14cu. in.) V-twin from the Paso but without the liquid cooling of the Paso. Ducati themselves saw the machine as an elegant café racer, which combined their handcrafted motorcycle-building skills with the very latest technology. This technology included an air/oil-cooled engine together with the company's unique and famous Desmodromic valve system mounted in a monoshock frame.

SPECIFICATION
Country of origin: ITALY
Capacity: 904cc (55.14cu. in.)
Engine cycle: 4-stroke
Number of cylinders: 2
Top speed: 120mph (193kph)
Power: 83bhp @ 7000rpm
Transmission: 6-speed
Frame: Tubular steel

BELOW: The Ducati 900 Supersport was a modern sports bike that combined 1990s technology with Ducati's traditions of handcrafted motorcycle construction.

ABOVE: The M600 Ducati Monster is powered
by an air-cooled, four-stroke, V-twin engine of
583cc (35.56cu. in.) displacement which
produces 53bhp @ 8250rpm. The engine and
main components are fixed to a skeletal frame.

E

ELITE

This German company was originally involved in car manufacture but merged with another German company, Diamant, in the early 1920s and began motorcycle production. Its first machines were powered by Kühne four-stroke engines of up to 500cc (30.50cu. in.) capacity. In the early years of the 1930s Elite produced motorcycles badged as EO models. These consisted of a Neander frame made under license from the Opel Motorclub and both 347 and 498cc (21.16 and 30.37cu. in.) overhead camshaft engines. The Opel frame was of pressed steel design. Later, the company used proprietary Sachs two-stroke and JAP four-stroke engines. World War II brought production to an end.

BELOW: Elite made motorcycles, such as this 1930 model, from a variety of bought-in components from a number of suppliers in the years prior to World War II. This overhead camshaft single uses a pressed steel frame.

ENFIELD INDIA

This company is based in Madras, India, and manufactures the Enfield Bullet in two capacities of 350 and 500cc (21.35 and 30.50cu. in.). The motorcycle they manufacture is actually a close derivative of the Royal Enfield Bullet that was introduced in Great Britain in 1949. The Royal Enfield company exported their machines all over the world, including India. The government there made arrangements to manufacture the Bullet during the 1950s which led to the formation of Enfield India Ltd. in 1955 and they have stayed in production ever since for both military and civilian customers. Refinements have only been introduced sparingly and when required by law. Items such as direction indicators have been fitted on export bikes, for example. In a complete reversal, the Indian-produced motorcycles are now exported to Britain as well as to many other countries.

1991 350 ENFIELD BULLET

The Indian-produced Enfield Bullet 350 is something of an anachronism in that it is a completely traditional design of 1950s'

British motorcycle that is still in production by historical accident. It features a swinging arm frame and also telescopic forks. The machine is powered by a single-cylinder engine. The Bullet is exported around the world where it is bought by those who

ABOVE: The Indian-made 1991 350 Bullet is essentially the same machine made in Britain by Royal Enfield in the 1950s although it had been slightly upgraded and featured improved electrics and turn signals.

desire a modern classic motorcycle, those who need simple and inexpensive transport, or by those who find it a curiosity.

SPECIFICATION
Country of origin: INDIA
Capacity: 346cc (21.10cu. in.)
Engine cycle: 4-stroke
Number of cylinders: 1
Top speed: 75mph (120kph)
Power: 18bhp @ 5625rpm
Transmission: 4-speed
Frame: Tubular steel

1993 ENFIELD BULLET

The 500cc Bullet is simply a larger capacity version of the 350 model and shares many cycle parts. It does have options such as a disc front brake. There are standard and deluxe versions available.

SPECIFICATION
Country of origin: INDIA
Capacity: 499cc (30.43cu. in.)
Engine cycle: 4-stroke
Number of cylinders: 1
Top speed: n/a
Power: 22bhp @ 5400rpm
Transmission: 4-speed
Frame: Tubular steel cradle

EXCELSIOR (USA)

The American company who chose the name Excelsior for their products diversified into motorcycle production in Chicago, Illinois, in 1907. They were already in business as the Excelsior Supply Company. The first motorcycle from this company was a 3.25hp single which had a displacement of 438.1cc (26.72cu. in.) and it was typical of

ABOVE: **The 500cc Enfield India Bullet. The main difference with the 350 is in the engine displacement achieved through a larger cylinder bore, as both models use the same cycle parts, accessories and finishes.**

ABOVE: **Several companies used the name Excelsior and this caused problems in export markets. American Excelsiors were sold as American-X in both single and twin form.**

the earliest days of internal combustion engine manufacture. The engine itself was mounted vertically in a frame which was clearly derived from a pedal bicycle, and had a mechanically-operated exhaust valve and an atmospheric inlet valve. This latter valve was opened by the vacuum created in the cylinder bore by the descending piston on the induction stroke. Both valves were mounted in a chamber alongside the cylinder but siamesed into it and this became known as a 'pocket-valve'.

Later the company produced larger capacity, single-cylinder machines which also used a pocket-valve design. Following the trend towards V-twins in American motorcycling, Excelsior dropped production of singles in 1913, a year after it had been taken over by Ignatz Schwinn, a bicycle manufacturer. Under Schwinn's management the company had considerable racing success in both board- and dirt-track events. They abruptly withdrew from racing

in 1920 after the death of their star rider, Bob Perry, in an accident at the Los Angeles Speedway while testing a new racer. Ignatz Schwinn is reputed to have smashed the castings of the race bikes with a sledgehammer immediately afterward. Excelsior stopped manufacturing motorcycles in 1931 after introducing a 740cc (45.14cu. in.) capacity V-twin that would eventually become a standard and popular capacity in American motorcycling.

1922 EXCELSIOR V-TWIN

Excelsior introduced their first V-twin in 1910 in line with the other American motorcycle manufacturers. The trend for V-twins in American motorcycling started early and by 1922 the machines featured an inlet-over-exhaust valve configuration V-twin in a tubular frame with trailing link forks and a rigid rear.

SPECIFICATION
Country of origin: USA
Capacity: 992cc (60.51cu. in.)
Engine cycle: 4-stroke
Number of cylinders: 2
Top speed: 95mph (152kph)
Power: n/a
Transmission: 3-speed
Frame: Tubular steel

EXCELSIOR (UK)

In England, a company known as Bayliss, Thomas and Company built Excelsior bicycles and later offered for sale bicycles fitted with Minerva engines. They developed this machine and by 1902 were offering a belt-drive machine with a 2.75hp single-cylinder engine fitted inclined forward under the front down-tube of the frame. The fuel and oil tanks were located within the central part of the diamond-shaped frame. The company was taken over by R. Walker & Son in 1919. Like many other manufacturers, Walker produced motorcycles with proprietary engines from companies such as Blackburne and JAP. While many of their machines were utility motorcycles, the company did build some race bikes and campaigned them at Brooklands and in the Isle of Man TT races. They won a lightweight event in 1929 and went on to debut an innovative racer at the 1933 event.

For the 1930s the company produced a range of bikes with Villiers engines fitted including numerically designated models of 147, 196 and 247cc displacement (8.96, 11.95 and 15.06cu. in. respectively). Larger capacity models had 245 and 300cc (14.94 and 18.30cu. in.) overhead-valve JAP engines. The 1933 Isle of Man bike became known as the 'Mechanical Marvel'. Its single-cylinder engine had a displacement of

ABOVE: The Depression meant that times were hard for British motorcycle manufacturers during the 1920s and 1930s so that companies such as Excelsior had to build whatever they could sell, like this medium capacity machine.

250cc (15.25cu. in.) and featured four radial valves in the cylinder head. They were complex but operated by a pair of camshafts arranged fore and aft of the cylinder. It was fast enough to win the TT Race but a difficult machine to tune. The range was upgraded and the models given names such as Manxman and Norseman but essentially the company saw out the decade manufacturing two- and four-stroke motorcycles. An unusual motorcycle was the Viking which was a fully enclosed machine fitted with a water-cooled engine. One prewar lightweight was the Autobyk, an autocycle which was fitted with a Villiers Junior engine and pedaling gear but no legshields. In the postwar period the model was reintroduced in standard and deluxe versions. Both were made until 1950.

During World War II, Excelsior produced the Welbike and then a number of other lightweight machines in the postwar years. These included a couple of autocycles, a Skutabyk, a scooter, and motorcycles with names such as Consort, Roadmaster and Talisman, many of which used Villiers engines. The company stopped producing motorcycles in 1965. The Skutabyk was introduced in 1957 and was an enclosed machine, a combination of scooter and motorcycle as hinted at by its name.

1926 EXCELSIOR 350 SPORTS

The majority of machines built by Excelsior were utility models but such was the nature of the motorcycle business in the 1920s that the company would build anything it thought it could sell. Their bikes were cam-

paigned in competitive events and after the 350 Sports had been built the company went on to a lightweight TT victory in 1929 with a 250cc race bike.

SPECIFICATION
Country of origin: GREAT BRITAIN
Capacity: 349cc (21.28cu. in.)
Engine cycle: 4-stroke
Number of cylinders: 1
Top speed: n/a
Power: n/a
Transmission: 4-speed
Frame: Tubular steel

1935 EXCELSIOR MANXMAN

The Manxman was an attempt by Excelsior to lift their name above all the other com-

BELOW: An Excelsior Manxman engine fitted to a later frame and forks and prepared for classic motorcycle racing.

panies who offered proprietary-engined motorcycles. It was launched in 1935 in two versions – 246 and 349cc (15 and 21.28cu. in.). They had the same stroke so the different displacement was made up by the dimensions of the bore. The engines were single-cylinder with a shaft-driven single overhead camshaft and the mag-dyno was gear-driven. The motorcycle was clearly designed for racing although, despite its name, it never actually won an Isle of Man TT race.

SPECIFICATION
Country of origin: GREAT BRITAIN
Capacity: 349cc (21.28cu. in.)
Engine cycle: 4-stroke
Number of cylinders: 1
Top speed: 85mph (140kph)
Power: 22bhp @ 6000rpm
Transmission: 4-speed
Frame: Cradle

EXCELSIOR (Germany)

Two companies in Germany used the Excelsior brandname: one between the years 1901 and 1939 at Brandenburg; and another briefly between 1923 and 1924 in Munich. The latter company only ever produced a two-stroke lightweight. The older company, owned by Conrad and Patz, produced motorcycles using Minerva, Zedel and Fafnir engines and later English JAP engines. Later they built bikes with Bark two-stroke engines which were made in Dresden. The fact that several companies used the same brandname caused problems when it came to exports: British Excelsiors when they were sold in Germany were renamed as Bayliss Thomas after the company's founders; and American Excelsiors were called American X when they were sold in England.

F

FAGAN

This was a short-lived Irish marque which was based in Dublin. Fagan, who were in business for a few years in the mid-1930s, assembled British-manufactured components as a way of getting around a tariff on imported vehicles. The Fagan was powered by a 148cc (9.02cu. in.) Villiers engine, and had a three-speed Albion gearbox, a Diamond Motors frame, Webb forks and a Lycett saddle. Few were sold so production was soon halted.

FANTIC

Fantic Motor was formed in 1968 and produces sporting mopeds and lightweight motorcycles in Barzargo in northern Italy. Fantic Motor uses two-stroke Minarelli engines in its motorcycles. They have been successful in the manufacture of off-road machines for trials competition, such as the Trial 307 and K-R00 models, and enduro-style machines such as the Caballero and Oasis models.

1977 FANTIC CHOPPER TX 131

There are numerous motorcycles described as 'factory customs' and the seemingly contradictory term has been widely accepted, as have motorcycles styled that way. One of the earliest and one which was not particularly successful was the Fantic TX 131 Chopper. It was clearly styled on the long-forked American Choppers, such as those which were seen in the film *Easy Rider*, but was considerably smaller and powered by a tiny two-stroke engine.

SPECIFICATION
Country of origin: ITALY
Capacity: 123.5cc (7.53cu. in.)
Engine cycle: 2-stroke
Number of cylinders: 1
Top speed: 74.6mph (120kph)
Power: 15bhp @ 7500rpm
Transmission: 5-speed
Frame: Tubular steel

1991 FANTIC K-R00

The Fantic K-R00 is one of the modern generation of specialized machines for off-road trials. The motorcycle features high ground clearance, a short wheelbase and is light in weight. These attributes make it a competitive mount. The engine is liquid-cooled and uses a refrigeration unit for cooling. It is of a light alloy construction, where possible, including the rear monoshock suspension

ABOVE: The 1991 Fantic K-R00 is a particularly specialized trials machine developed especially for off-road competition. The model features disc brakes and a liquid-cooled engine.

system. The forks are upside-down telescopics and braking is by means of discs, both at the front and at the rear.

SPECIFICATION
Country of origin: ITALY
Capacity: 249.4cc (15.21cu. in.)
Engine cycle: 2-stroke
Number of cylinders: 1
Top speed: 61.5mph (99kph)
Power: 21bhp @ 5250rpm
Transmission: 6-speed
Frame: Chrome molybdenum cradle

FLANDRIA

This Belgian company was originally a small family firm run by A. Claeys Flandria that produced agricultural implements and bicycles. The company began motorcycle production in 1950 based at Zedelgem and Zweevezele in Belgium. By the mid-1970s the company was producing in excess of 100,000 motorcycles and mopeds each year, alongside extruded aluminum products, lawnmowers and heating equipment. They have overseas construction plants in the Netherlands, France, Morocco and Portugal, and have their engines license-built in Figuras, Spain.

BELOW: The 1975 SP 947 MS is a sports moped typical of the products from Belgian manufacturer Flandria. It is powered by a two-stroke engine.

1975 SP 947 MS

This model is typical of the range of machines offered by Flandria, all of which are based around small capacity two-stroke engines. There are a variety of styles including an off-road machine known as the Scorpion M and a less sporting motorcycle called the Rekord 6.

SPECIFICATION
Country of origin: BELGIUM
Capacity: 49.7cc (3.03cu. in.)
Engine cycle: 2-stroke
Number of cylinders: 1
Top speed: 52.8mph (85kph)
Power: 6bhp @ 7800rpm
Transmission: 4-speed
Frame: Tubular cradle

ABOVE: The Flying Merkel models were popular racing motorcycles. This photograph shows factory rider Maldwyn Jones with his machine after a race in 1913.

FLYING MERKEL

This was another relatively short-lived American brand of motorcycle that was produced between 1909 and 1917. The concern was initially based in Milwaukee, Wisconsin, but moved to a new factory in Middletown, Ohio, in 1911. It was the amalgamation of Joe Merkel's company that started in Milwaukee, Wisconsin, in 1902, the Light Manufacturing and Foundry Company of Pottstown, Pennsylvania, and the Miami Cycle Company.

FN

This was a Belgian company that was amongst the pioneers in European motorcycle manufacture. The FN company used innovative engineering in order to produce in-line four-cylinder machines and shaft-drive motorcycles as early as 1903. The shaft-drive feature endured until 1923 and the company became famous for success in racing and long distance events. The company has also manufactured weapons including the FN rifle. The company's earliest machines featured 225 and 286cc (13.72 and 17.44cu. in.) single-cylinder engines and 496 and 748cc (30.25 and 45.62cu. in.) four-cylinders. In 1924 the company produced chain-drive motorcycles because of considerably lower manufacturing costs and made 348 and 498cc (21.22 and 30.37cu. in.) displacement machines. During the period of the 1930s, a 198cc

(12.07cu. in.) Villiers-engined machine was in FN's range and later in that decade a flat-four was produced for the Belgian Army. Alongside this, the company had some success with racing singles designed by Dougal Marchant. One such bike was of 348cc (21.22cu. in.) displacement and the other of 498cc (30.37cu. in.). From these was developed a supercharged 498cc (30.37cu. in.) overhead camshaft vertical-twin.

In the postwar years FN produced a range of unit construction twins in side-valve and overhead-valve forms, as well as some small capacity two-strokes. Certain of these motorcycles were fitted with an unusual trailing link design of forks. With these machines the company achieved some success in various motocross events. Production of motorcycles stopped in 1957.

1904 3HP FN

The 1904 3hp FN was in some ways technologically advanced for its time in that it

ABOVE: The FN Four, introduced in 1904, was the first successful four-cylinder motorcycle ever produced. It featured shaft drive.

was one of the first four-cylinder machines which was manufactured with shaft drive. In other ways, however, it was typical of its era, in that it was simply a bicycle frame with an engine fitted.

SPECIFICATION
Country of origin: BELGIUM
Capacity: 496cc (30.25cu. in.)
Engine cycle: 4-stroke
Number of cylinders: 4
Top speed: 40mph (64kph)
Power: n/a
Transmission: Single-speed
Frame: Diamond

FRANCIS–BARNETT

This company came about in 1919 through the collaboration of Arthur Barnett and Gordon Francis. Barnett was already producing motorcycles in Coventry, England. These motorcycles were known as Invicta and used Villiers and JAP engines. The first machine from the new company, with a bolted tubular frame and a JAP engine, was introduced in 1920. Francis–Barnett manufactured machines with triangulated frames from 1923 onward and became noted for the production of lightweight machines. These machines were powered by Villiers

BELOW: The 1935 Francis–Barnett Plover 41 featured a frame with an unusual frame front down-tube made from twin pressings. In all other aspects it was of traditional design.

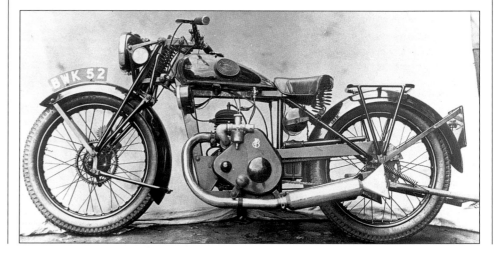

engines of 147, 172 and 196cc (8.96, 10.49 and 11.95cu. in.) displacement.

After World War II was over, the company returned to the motorcycle market with an autocycle – the Powerbike 50 – and a lightweight motorcycle – the Model 51 Merlin. As in the prewar days, this was a Villiers-engined motorcycle. In 1947 the company became part of the AMC group who produced Matchless and AJS machines. The AMC group also acquired James and in later years the two companies' motorcycles were the same except for the tank badges. The company ceased production of motorcycles in 1966.

1928 MODEL 10 PULLMAN

The Model 10 Pullman bike was Francis–Barnett's attempt to offer a luxury motorcycle. It was not a notable success largely because its extra cost far outweighed its luxury features. The Villiers engine was mounted lengthways into the frame, which was of an unusual appearance in order to accommodate it. Also of unusual appearance was the fuel tank – a triangular item fitted between the frame tubes. The frame was of a simple design in that six pairs of straight tubes, one bent pair and a steering head were bolted together to complete the frame. While the Pullman was discontinued in 1928, certain of the features of its construction survived in other Francis–Barnett products, most notably in the Model 16 Dominion.

SPECIFICATION
Country of origin: GREAT BRITAIN
Capacity: 343cc (20.92cu. in.)
Engine cycle: 2-stroke
Number of cylinders: 1
Top speed: 60mph (96.5kph)
Power: n/a
Transmission: 3-speed
Frame: Bolted tubular

ABOVE: Launched by Francis–Barnett in 1933, the Cruiser was unusual in that it featured an I-section forging as the frame down-tube to which were added channel sections.

1931 MODEL 19 147

The Model 19 was reintroduced by Francis–Barnett in 1931 as a 147cc (8.96cu. in.) displacement motorcycle. It was not supplied with lights, whereas the otherwise identical Model 20 was. For the following year the bikes were renamed the 23 Merlin and 24 Kestrel as Francis–Barnett started a policy of both numbering and naming their bikes after birds. The Plover, for example, was a 1936 Model.

SPECIFICATION
Country of origin: GREAT BRITAIN
Capacity: 147cc (8.96cu. in.)
Engine cycle: 2-stroke
Number of cylinders: 1
Top speed: n/a
Power: n/a
Transmission: 2-speed
Frame: Bolted tubular

ABOVE: The triangular frame seen on this 1931 Model 19 147 was typical of a frame type that Francis–Barnett pioneered and adhered to.

1937 G45 CRUISER

The Cruiser was a new model from Francis–Barnett in 1933 and it remained in their range for several years. The 1937 model, in the same way as its predecessors, was fitted with a Villiers 249cc (15.18cu. in.) two-stroke engine with a bore and stroke of 63 and 80 mm. The Cruiser, in this guise, lasted until the outbreak of World War II.

SPECIFICATION
Country of origin: GREAT BRITAIN
Capacity: 249cc (15.18cu. in.)
Engine cycle: 2-stroke
Number of cylinders: 1
Top speed: n/a
Power: n/a
Transmission: 4-speed
Frame: Tubular steel

ABOVE: The G45 Cruiser was produced from 1933 to the outbreak of World War II. It used a Villiers two-stroke engine.

1962 ISDT 250CC

The International Six Days' Trial (ISDT) has always been a prestigious event and manufacturers prepared motorcycles especially for the teams competing in it. By 1962 Francis–Barnett had long been part of AMC who were producers of the AJS and Matchless marques, both of which had pedigree in trials and scrambles. The company produced trials bikes with Villiers and then an unpopular two-stroke AMC engine. They persevered with this engine into the 1960s, the decade which saw the end of their motorcycle production.

SPECIFICATION
Country of origin: GREAT BRITAIN
Capacity: 249cc (15.18cu. in.)
Engine cycle: 2-stroke
Number of cylinders: 1
Top speed: n/a
Power: n/a
Transmission: 4-speed
Frame: Tubular steel

ABOVE: The trials models made by Francis–Barnett from 1959 onward used the 249cc (15.18cu. in.) AMC single-cylinder engine.

G

GARELLI

This company was founded by Alberto Garelli in Italy in 1913. Garelli designed an unusual two-stroke engine that featured two pistons in a single-cylinder. It was progressively refined between 1913 and 1935.

BELOW: This Garelli race bike features the double-piston two-stroke engine designed in 1913. It was further developed up until 1935.

During the 1920s riders on such machines had considerable racing success and broke a number of records. In the postwar years the company produced the 38cc (2.31cu. in.) Mosquito engine for use in conjunction with bicycles. In 1967 Garelli merged with the Agrati organization. Two-stroke engines are produced in a factory in Sesto San Giovanni and another factory at Monticello assembles a range of Garelli mopeds and light motorcycles.

BOTTOM: An Agrati-manufactured Garelli. This water-cooled, twin-cylinder 125 racer from the 1981-2 season has lightweight components.

1981 GARELLI KATIA MOPED

The Katia was a popular step-through moped manufactured during the 1970s and 1980s. It was one of their range of small capacity machines that included sports motorcycles and mopeds as well as basic transportation such as the Katia. The example seen here was a limited edition built especially for the famous London, England, Earl's Court Motorcycle Show. It featured an all-black paint scheme with magnesium alloy wheels and new moulded legshields in order to dress up the horizontal single-cylinder machine.

SPECIFICATION
Country of origin: ITALY
Capacity: 46cc (2.80cu. in.)
Engine cycle: 2-stroke
Number of cylinders: 1
Top speed: 25mph (40kph)
Power: 1.5bhp @ 4000rpm
Transmission: 2-speed
Frame: Step-through

ABOVE: The step-through design of moped has numerous advantages, especially for women riders who want to ride wearing skirts. The 1981 Garelli Katia shown here was produced as cheap, functional transportation.

GAS GAS

This Spanish company was founded in 1986 but its origins go further back, to 1982 and the Merlin marque. This latter company was founded in part by Ignacio Bulto, the son of Francisco Bulto who founded Bultaco in 1958. The Merlin Company specialized in trials bikes powered by 347cc (21.16cu. in.) Cagiva two-stroke engines and ceased production in 1986. From this emerged the Gas Gas operation which also specializes in off-road machines. These are in the main for trials and enduro use and utilize their own engines as well as those from the Italian companies of Cagiva and TM. TM made its own mopeds, lightweight motorcycles and off-road bikes between 1968 and 1992 as well as selling its engines on to other motorcycle manufacturers.

GILERA

The first motorcycles from this company were designed by Giuseppe Gilera in 1909. Initially 317cc (19.33cu. in.), belt-driven, overhead-valve engines were utilized in a diamond frame. The switch to side-valve engines occurred soon after and lasted until 1926 when Gilera again offered ohv-engined bikes. In 1935 Gilera bought the Rondine – a machine made by an aeronautical company who up until that time were producing a very successful supercharged four-cylinder motorcycle engine that featured inclined cylinders, double overhead camshafts and water cooling. This acquisition became the basis of Gilera's race bikes and was sequentially upgraded and refined, lasting until the mid-1960s.

In 1969 Gilera was taken over by the Piaggio group who manufacture Vespa scooters. Under this new ownership Gilera introduced a range of small capacity bikes including the 200T4, 125TGI and 50TS. These designations approximated the displacement of each machine in cubic centimetres. The 200T4 was a four-stroke while the smaller machines were two-strokes. Larger capacity bikes from this company reappeared during the 1980s when the company also produced a range of step-through mopeds.

ABOVE: The Gas Gas company manufacture specialized trials and enduro machines for off-road competition. This model is designed with high ground clearance and low seat, light weight, disc brakes and a liquid-cooled, two-stroke engine.

BELOW: The postwar ban on superchargers for racing bikes forced Gilera to build a new 498cc (30.37cu. in.) machine. The naturally-aspirated engine was a double overhead camshaft, air-cooled, four-cylinder unit designed by Pietro Remor. The 1956 model is shown.

ABOVE: The Gilera NordWest of 1993 is based around Gilera's single-cylinder engine and, although primarily designed for road use, has the look of the desert 'raid' bikes used in events such as the Paris-Dakar.

1956 GILERA WORKS RACER

The ban on forced induction, i.e. super-charged racers, after World War II forced Gilera to redesign their racing machines. Pietro Remor designed double overhead camshaft, air-cooled, four-cylinder machines which in developed form went on to win numerous races with a variety of the sport's top riders in the saddle.

SPECIFICATION
Country of origin: ITALY
Capacity: 498cc (30.37cu. in.)
Engine cycle: 4-stroke
Number of cylinders: 4
Top speed: n/a
Power: n/a
Transmission: 4-speed
Frame: Duplex

1993 GILERA NORDWEST

The NordWest has the look of the big Paris-Dakar enduro bikes but is intended to be a roadgoing machine and was fitted with road tires accordingly. It was based around

RIGHT: The Gilera Saturno 500 of 1991 is powered by a four-stroke, liquid-cooled, double overhead camshaft, four-valve, single-cylinder engine. It is primarily a sports bike and is fitted with a half-fairing and monoshock rear suspension around a steel space frame.

Gilera's large displacement single-cylinder engine. In 1985 the engine appeared with a displacement of 350cc (21.35cu. in.) and was used in enduro machines such as the Gilera Dakota. However, the engine was later increased in capacity to 558cc (34.03cu. in.) and utilized in the NordWest and the Nuovo Saturno café racer.

SPECIFICATION
Country of origin: ITALY
Capacity: 558cc (34.03cu. in.)
Engine cycle: 4-stroke
Number of cylinders: 1
Top speed: 116mph (187kph)
Power: 52bhp @ 6500rpm
Transmission: 5-speed
Frame: Aluminum

GNOME ET RHONE

This French concern started out as an aircraft engine manufacturer during World War I and started producing motorcycles in 1919. They began by building British-designed motorcycles under license. From the early 1920s they produced machines of their own designs. A range of different displacement engines of both side- and overhead-valve configurations were produced and then later on a number of flat-twins mounted in pressed steel frames. These BMW-style machines were available with both 495 and 745cc (30.19 and 45.44cu. in.) engines. In the last years of the decade before World War II, Gnome et Rhone had six motorcycles in production: The Junior, Major and Super Major, as well as the D5, CV2 and Type X. The first three were 250cc (15.25cu. in.) and 350cc (21.35cu. in.) four-speed machines while the latter three were a 500cc (30.50cu. in.) single, a 500cc (30.50cu. in.) twin and a 750cc (45.75cu. in.) flat-twin. The flat-twin was enlarged to 800cc (48.80cu. in.) when produced for the French Army prior to World War II.

The company resumed production after the war and continued making motorcycles – mostly two-strokes of less than 200cc (12.20cu. in.) – until 1959.

GREEVES

The company was named after Bert Greeves who only came into the motorcycle industry because he had a disabled cousin. He fitted an engine to his invalid cousin's wheelchair and the pair started the manufacture of invalid carriages. The company flourished after World War II and the duo intended to begin to manufacture motorcycles. They embarked on a development plan which included entering a prototype in

ABOVE: The 1963 Greeves 247 TS was powered by a 247cc (15.06cu. in.) two-stroke engine as might be expected from its designation. It was unusual in its frame design which featured a cast alloy beam in place of front down-tubes. The front suspension used rubber bushes.

competitive off-road events. This meant that the machine became known and allowed both spectators and competitors to see its advanced, albeit unusual, design features. The Greeves used rubber bushes as its suspension as did their invalid carriages. The engine was a 197cc (12cu. in.) Villiers two-stroke which was fitted into a duplex loop frame. Drum brakes and a dual seat completed the unorthodox motorcycle. In 1953 the first production machines were assembled using a unique frame that included a cast alloy beam. There were two roadgoing models, the three-speed 20R and the deluxe four-speed 20D. Later the company built a machine that utilized a British Anzani two-stroke twin engine. In the mid-1950s Greeves dropped their rubber rear suspension set-up in favor of the more conventional hydraulic dampers. At the end of that decade the company mar-

keted models named after famous trials, namely the Scottish and the Hawkstone, which reflected the success of the Greeves marque off-road. In the 1960s the company produced 250cc road-racing motorcycles and its own design of engine. Production stopped in 1972 mainly as a result of imported competing makes and the retirement of Bert Greeves.

1963 GREEVES 247 TS

Greeves motorcycles had a somewhat unusual appearance through the use of a

rubber torsion suspension system. Early off-road bikes featured this both front and rear but the rear suspension was changed to a more conventional hydraulically-damped set-up in 1956. Greeves used a cast alloy beam in place of front down-tubes and fitted a 247cc (15.06cu. in.) Villiers engine. With this design their riders, including Dave Bickers and Brian Stonebridge, dominated the 250cc class in motocross for several years.

SPECIFICATION
Country of origin: GREAT BRITAIN
Capacity: 247cc (15.06cu. in.)
Engine cycle: 2-stroke
Number of cylinders: 1
Top speed: 60mph (96.5kph)
Power: n/a
Transmission: 4-speed
Frame: Cast alloy beam

H

HARLEY–DAVIDSON

The legendary American Harley–Davidson motorcycle company was founded in 1903 in Milwaukee, Wisconsin, by William Harley, the Davidson Brothers – Walter and Arthur – and their father, William. The men worked at evenings and weekends to produce their earliest machines. A shed out in William Davidson's backyard became their first factory and in 1905 Walter Davidson became the first full-time employee of the fledgling concern. Their 1905 motorcycles were singles of 405cc (24.70cu. in.) displacement in a bicycle-style frame with belt drive and no brakes. The rider pedaled backward in order to slow down.

The Silent Gray Fellow was the first model produced in significant numbers by the workforce that now numbered six. It was still a single and so named because of its quiet running and optional gray paint scheme. In 1907 William Davidson also went to work for the company full-time and the three formed a corporation and sold shares. The two families retained a controlling portion of the stock right up until the company was sold to American Machine and Foundry in 1969. Harley's first V-twin was introduced in 1909 but its first really successful one was marketed in 1911. A chain drive version appeared in 1912 and in 1913 Harley–Davidson sold 12,904 motorcycles and export sales began. The foundations for the company as it is known today had been laid.

During World War I Harley–Davidson supplied half its output of motorcycles to the US Army and prepared for postwar production including expansion at the factory. The company suffered a low in sales in the early 1920s as a result of the competition from low-priced automobiles but retained a strong export market. The Wall Street Crash and its effects devastated Harley's sales until they bottomed out in 1933 with sales of only 3,703 bikes. In 1936 they introduced the EL Knucklehead – their first overhead-valve engined motorcycle – which was an instant success. Unionization of the workforce occurred in 1937 and the first of the founders died in the same year. The others would pass away in 1942, 1943 and 1950. World War II interrupted the production of civilian motorcycles and Harley–Davidson manufactured in excess of 88,000 side-valve 740cc (45.14cu. in.) motorcycles for the various Allied armies. These bikes later popularized Harley–Davidsons around the world, particularly in parts of Europe.

Postwar, once production was back to normal, the Panhead engine appeared and

Harley enjoyed boom years, despite competition from huge numbers of imported British motorcycles. By the mid-1960s their US market share had contracted considerably and as little as 3 percent of production was exported. The future was uncertain and in 1968 American Machine and Foundry, a huge conglomerate, was one of two companies interested in a takeover. They took control in January 1969.

The years that followed were something of a mixed blessing for Harley–Davidson – productivity increased but quality control decreased to the extent that AMF bikes did not gain a good reputation. The company also had to face another wave of imports directed at its US market – those from Japan. Despite a number of new products, by 1980 AMF was looking at ways of disposing of Harley–Davidson and accepted a management buyout plan by 13 of Harley–Davidson's executives. The biggest news, besides the change of ownership, was the first all-new Big Twin engine since 1965 – the Evolution engine – for which much of the development work had been carried out under AMF. The Evolution engine restored Harley–Davidson to the road to success and undoubtedly made it the strong company it has remained ever since.

During their history the company has experimented with a number of smaller

BELOW: The Electra Glide is the epitome of the Harley–Davidson as an American icon. As a model it has been in production for a period of more than 30 years.

capacity motorcycles including, in the postwar years, the Hummer, which was a 125cc (7.62cu. in.) DKW-inspired machine, the Aermacchi lightweights in the 1960s, and the successful Sportster range which has been part of their catalog since the K Model of 1952.

1906 HARLEY–DAVIDSON SILENT GRAY FELLOW

The Silent Gray Fellow was so named because of its quiet running (as a result of its muffler) and the Renault gray paint scheme. The engine size was sequentially increased from 405cc (24.70cu. in.) when introduced to 500cc (30.50cu. in.) in 1909 and then to 565cc (34.46cu. in.) in 1913. Other changes made in the duration of the production run were to the design of the cylinder-head cooling fins and a reshaping of the front down-tube of the frame. The

BELOW: The Silent Gray Fellow was so named as a result of the quiet running of its engine and also due to its gray-painted components.

gas tank was redesigned in 1912 and 1916 while belt drive was discontinued in 1914. Production of all singles ended in 1918 with the trend toward V-twins, which were seen as a way in which to increase the power of a motorcycle engine fairly cheaply. Also the design fitted existing frames.

SPECIFICATION
Country of origin: USA
Capacity: 405cc (24.70cu. in.)
Engine cycle: 4-stroke
Number of cylinders: 1
Top speed: 50mph (80kph)
Power: n/a
Transmission: Single-speed
Frame: Steel loop

1930 HARLEY–DAVIDSON VL MODEL

The V Model series was an almost entirely new motorcycle when it was introduced sharing few parts with its F-head predecessor. Around 13 variants of the side-valve V Models were produced through the production run that had variations in specifications such as those equipped with magnetos and higher compression bikes. The V Series became the U Series in 1937 with the introduction of dry sump lubrication.

BELOW: **The large displacement V-twins were manufactured as V and VL models between 1930 and 1937. They were redesignated the U and UL models in the latter year.**

SPECIFICATION
Country of origin: USA
Capacity: 1207cc (73.62cu. in.)
Engine cycle: 4-stroke
Number of cylinders: 2
Top speed: 85mph (140kph)
Power: n/a
Transmission: 3-speed
Frame: Steel loop

1936 HARLEY–DAVIDSON 61 OHV

The 'Knucklehead', as this model became known because of the shape of its rocker covers, was the first Harley–Davidson overhead-valve V-twin and it featured dry sump lubrication where the oil circulated between a tank and the engine instead of the earlier total loss systems. This motorcycle set the style for Harley–Davidsons that still endures with its rounded two-halves gas tank with the speedo set into it.

SPECIFICATION
Country of origin: USA
Capacity: 988.56cc (60.30cu. in.)
Engine cycle: 4-stroke
Number of cylinders: 2
Top speed: 90mph (150kph)
Power: 40bhp
Transmission: 4-speed
Frame: Double loop

BELOW: **The Knucklehead was the first overhead-valve Harley–Davidson big twin. It was introduced in 1936 but production was interrupted by the outbreak of World War II.**

1942 HARLEY–DAVIDSON WLA

The prewar side-valve-engined motorcycle was militarized and supplied to Allied armies during World War II. There were two variants, the WLA and WLC. The WLA was an WL with an A suffix for Army that was supplied to both American and Chinese Armies, while the WLC was a WL with a C suffix. It was originally manufactured for the Canadian forces but it was also supplied to the forces of Britain, Russia, Australia and South Africa. Harley built in excess of 88,000 machines for the war effort and earned a special award from the US Army for doing so. It is widely believed that these machines helped to popularize Harley–Davidson motorcycles in new export markets in the years after World War II.

SPECIFICATION
Country of origin: USA
Capacity: 739.38cc (45.12cu. in.)
Engine cycle: 4-stroke
Number of cylinders: 2
Top speed: 65mph (104kph)
Power: 23bhp @ 4600rpm
Transmission: 3-speed
Frame: Steel loop

BELOW: The 740cc (45.14cu. in.) WL Series became the WLA for the duration of World War II. Along with a Canadian variant, the WLC, approximately 88,000 were made for the Allied armies from 1939 to 1945.

1942 HARLEY–DAVIDSON XA

This was an experimental motorcycle built at the request of the US Government and closely styled on the flat-twin German BMW machines in use by the Wehrmacht at the time. Despite its acceptance as a practical motorcycle, the US Army did not order vast numbers for its war effort. The reasons for this were that by the time it was ready for mass production the army's requirement had changed due to the success of the Jeep and also because of the number of WLA and WLC models which had already been produced for America and her allies.

SPECIFICATION
Country of origin: USA
Capacity: 738cc (45.01cu. in.)
Engine cycle: 4-stroke
Number of cylinders: 2
Top speed: n/a
Power: 23bhp @ 4600rpm
Transmission: 4-speed
Frame: Tubular steel plunger

BELOW: The XA was an experimental bike from Harley–Davidson, built at the request of the US Government. Based on German designs in use by the Wehrmacht, it was never more than experimental and only 1000 were ever made.

1949 HARLEY–DAVIDSON FL HYDRA-GLIDE

Harley–Davidson introduced their Panhead engine in 1948. It was the successor to the overhead-valve Knucklehead, also having overhead-valves. The new engine was originally fitted to Harley's springer fork equipped rigid frame rolling chassis but this later benefited from hydraulic telescopic forks and so became designated the Hydra-Glide. Later, when a swingarm frame was used to give the Panhead front and rear suspension, it was redesignated the Duo-Glide, to indicate two suspension systems.

BELOW: The FL was updated in 1948 with the introduction of the Panhead engine. It was further upgraded the following year when telescopic forks were fitted, turning it into the FL Hydra-Glide.

In 1965, an electric starter became available, and it was renamed the Electra-Glide.

SPECIFICATION
Country of origin: USA
Capacity: 1200cc (73.20cu. in.)
Engine cycle: 4-stroke
Number of cylinders: 2
Top speed: 102mph (164kph)
Power: 55bhp
Transmission: 4-speed
Frame: Tubular steel

1950 HARLEY–DAVIDSON MODEL G SERVICAR

The three-wheeled Servicar was introduced by Harley–Davidson in 1932. It was aimed at commercial users such as garages and delivery agencies and soon found favor with police traffic departments in expanding urban areas. Due to its usefulness, the Model G Servicar stayed in production until 1973. During this long production run the machine was upgraded only sparingly. Every Servicar was powered by the 740cc (45.14cu. in.) flat-head V-twin engine but the rear load box, originally made from steel, was later manufactured in glassfiber.

SPECIFICATION
Country of origin: USA
Capacity: 740cc (45.14cu. in.)
Engine cycle: 4-stroke
Number of cylinders: 2
Top speed: 50mph (80kph)
Power: 23bhp @ 4600rpm
Transmission: 3-speed
Frame: Tubular steel

1971 HARLEY–DAVIDSON FX SUPERGLIDE

The Superglide was an early factory custom from Harley–Davidson, combining as it did the lighter Sportster front end with the brawn of the Big Twin Shovelhead engine. Designed by Willie G. Davidson, the grandson of one of the founders, it was seen as a watershed in US motorcycle development.

SPECIFICATION
Country of origin: USA
Capacity: 1207cc (73.62cu. in.)
Engine cycle: 4-stroke
Number of cylinders: 2
Top speed: 116mph (187kph)
Power: n/a
Transmission: 4-speed
Frame: Tubular steel

1977 HARLEY–DAVIDSON XLCR

The XLCR was a special limited edition version of the Sportster of which only 3000 were ever made. The XLCR was a café racer styled Sportster with a nose fairing and alloy wheels. The XLCR was based around a derivative of the race-proven XR-750 frame and was powered by a 997cc (60.81cu. in.) iron head Sportster engine.

SPECIFICATION
Country of origin: USA
Capacity: 997cc (60.81cu. in.)
Engine cycle: 4-stroke
Number of cylinders: 2
Top speed: 124mph (200kph)
Power: 61bhp @ 6200rpm
Transmission: 4-speed
Frame: Tubular steel

LEFT: The XLCR was a limited edition version of the Sportster model. Despite Harley–Davidson's race-track success, the café racer – hence the CR suffix – was not hugely popular. Today it is highly collectable.

1979 HARLEY–DAVIDSON FLH 1200 ELECTRA GLIDE

The name Electra Glide lost its hyphen down the years and gained weight as it featured larger saddlebags and fairings usually made from glassfiber, partially because of AMF's part ownership of a glassfiber manufacturing plant. The motorcycle in this form became famous to many people as *the* Harley–Davidson and it was immortalized on film in the movie *Electra Glide in Blue*, made in 1973. By 1979 the motorcycle had been powered by the 'Shovelhead' engine since 1966, and after 1984 would be powered by the Evolution Engine. This engine is of larger displacement at 1340cc (81.74cu. in.) and it powers Harley's biggest tourers like the FHTC Electra Glide Classic.

SPECIFICATION
Country of origin: USA
Capacity: 1207cc (73.62cu. in.)
Engine cycle: 4-stroke
Number of cylinders: 2
Top speed: 105mph (170 kph)
Power: 66bhp @ 5200rpm
Transmission: 4-speed
Frame: Tubular steel

RIGHT: The Electra Glide was introduced in 1965 in this form. Powered by a Panhead engine, it was a Duo-Glide, updated with an electric starter. The next upgrade would be the use of the Shovelhead engine.

BELOW: The Shovelhead-powered Electra Glide debuted in 1966 and stayed in production through the period of AMF's ownership of Harley–Davidson and into the company's return to private ownership. The model is now powered by the Evolution engine.

ABOVE: The FXWG was one of Harley–Davidson's so called 'factory customs'. It was an FX with a 'wide glide' front end, hence the WG suffix and a flame paint job typical of custom bikes of the time.

1981 HARLEY–DAVIDSON FXWG WIDE GLIDE

The Shovelhead V-twin engine powered one of the first factory custom bikes from Harley–Davidson in 1981. The FXWG was available with a 'flamed' paint scheme.

SPECIFICATION
Country of origin: USA
Capacity: 1340cc (81.74cu. in.)
Engine cycle: 4-stroke
Number of cylinders: 2
Top speed: 102mph (164kph)
Power: 49bhp @ 5800rpm
Transmission: 4-speed
Frame: Duplex tubular steel

1984 HARLEY–DAVIDSON XR 1000

The XR 1000 was an exercise in both engineering and marketing in that it was the closest street bike a customer could buy to a racing machine. The engine was derived from the XR-750 racer with a Sportster XL 1000 bottom end and a set of iron cylinders and alloy heads. Dual carburetors were fitted on the right of the engine and twin exhausts on the left. The remainder of the bike was the basic XLX Sportster.

BELOW: The XR1000 was a Sportster that reflected Harley–Davidson's racing successes in both flat- and dirt-track events. It was a street bike with the styling of competition machines and a tuned engine.

SPECIFICATION
Country of origin: USA
Capacity: 998cc (60.87cu. in.)
Engine cycle: 4-stroke
Number of cylinders: 2
Top speed: 120mph (193kph)
Power: 70bhp @ 6000rpm
Transmission: 4-speed
Frame: Tubular cradle

1986 HARLEY–DAVIDSON FXST SOFTAIL

The Softail was a restyled Big Twin from the Harley–Davidson factory. Its engine was the Evolution engine that had been introduced in 1983-4. It had first appeared in the FX models and spread across the complete range over the next three years. In traditional swingarm models the engine was

ABOVE: Harley–Davidson makes a much of its own heritage, appealing to the nostalgic-minded with models such as the Heritage Softail Nostalgia, which is a modern bike with vintage styling.

rubber-mounted to minimize vibration. The Softails take their styling from an earlier age and have the appearance of rigid frames as the suspension is hidden. The Softails feature a chrome horseshoe oil tank and traditional fenders. One model – the FXSTS – takes nostalgia one step further – it has springer forks. The Evolution and the Softail styling made Harley–Davidson motorcycles huge sellers in the 1980s and 1990s.

SPECIFICATION
Country of origin: USA
Capacity: 1340cc (81.74cu. in.)
Engine cycle: 4-stroke
Number of cylinders: 2
Top speed: 112mph (180kph)
Power: 67bhp
Transmission: 4-speed
Frame: Steel softail

1991 HARLEY–DAVIDSON FLSTF FAT BOY

The Fat Boy was heralded as a spectacular motorcycle when it was unveiled. It featured solid 16in. (400mm) diameter wheels back and front, and Evolution engine power in a Softail frame meaning the bike had a solid appearance.

BELOW: The Fat Boy is a version of Harley–Davidson's FL models. It is based around a Softail frame. It became a great success on its debut and the unusal name gave rise to the naming of Harley–Davidson's Bad Boy.

SPECIFICATION
County of origin: USA
Capacity: 1340cc (81.74cu. in.)
Engine cycle: 4-stroke
Number of cylinders: 2
Top speed: 115mph (185kph)
Power: 58bhp @ 5000rpm
Transmission: 5-speed
Frame: Steel softail

ABOVE: The 1340cc (81.74cu. in.) Harley–
Davidson Fat Boy is an unusually-named
motorcycle that combines traditional Big Twin
styling with modern components including
disc brakes and cast alloy wheels.

1991 HARLEY–DAVIDSON XLH SPORTSTER

While much of Harley's range is based around the 1340cc (81.74cu. in.) so-called Big Twin Evolution engine, the company have yet another range in production. The Sportster range is based around a unit construction engine. Two displacements are available: the 883 and 1200cc (53.86 and

ABOVE: The range of models referred to as Sportsters, such as this 1989 883 model, are based around two displacements of unit construction engines.

BELOW: The Bad Boy is a special FX model Harley–Davidson. It is based around the FX Softail Springer, hence its designation of FXSTSB. The bike is considered to be a contemporary but nostalgic factory custom.

73.20cu. in.) versions. The 883cc version is seen by many as an entry-level Harley but the fact that there are specific race series for this model indicate otherwise.

SPECIFICATION
Country of origin: USA
Capacity: 883cc (53.86cu. in.)
Engine cycle: 4-stroke
Number of cylinders: 2
Top speed: 110mph (178kph)
Power: 49bhp @ 6000rpm
Transmission: 5-speed
Frame: Tubular steel

1995 HARLEY–DAVIDSON FXSTSB BAD BOY

The FXSTSB Bad Boy is Harley's most nostalgic motorcycle to date. It features a Softail frame that has the old Harley look about it. It also has springer forks which, until the FXSTS, had not been fitted since 1948, and a hot-rod paint scheme. It has up-to-date features too, including the Evolution engine and disc brakes.

SPECIFICATION
Country of origin: USA
Capacity: 1340cc (81.74cu. in.)
Engine cycle: 4-stroke
Number of cylinders: 2
Top speed: 120mph (193kph)
Power: 58bhp @ 5000rpm
Transmission: 5-speed
Frame: Softail cradle

HEDLUND

Nils Hedlund founded this company in 1955 and manufactured Albin engines used in Swedish motocross bikes that had success during the 1950s and 1960s. He manufactured a number of complete motorcycles that were known as NH (his initials). The company stopped producing motorcycles in 1987 but before that, in the mid-1970s, the company manufactured a four-stroke V-twin of 1000cc (61cu. in.) displacement for use in motocross outfits. It was of unorthodox appearance, the tall cylinders having a square external section and the crankcases also being angular.

HENDERSON

This American Company was founded in Detroit by the Henderson brothers, Will and Tom, who came from an automotive background, since their father manufactured Henderson cars. The first Henderson motorcycle that went into production was made as a prototype in 1911 and featured an in-line four-cylinder engine, fitted into what appeared as an elongated bicycle-type frame. The machine was equipped with a

ABOVE: The Henderson company was founded in 1911 and was amongst those which pioneered four-cylinder motorcycle engines. They persevered with this configuration, as shown on this 1925 model.

long cylindrical fuel tank that ran horizontally between the frame tubes. The engine functions of the first model were controlled by a handlebar twistgrip throttle, a hand-clutch and a pair of foot-pedals. In 1917 the company was purchased by bicycle manufacturer Ignatz Schwinn who already owned Excelsior. Will Henderson left Schwinn's employ sometime later and went to work for the Ace company. Ace was later acquired by Indian Motorcycles who continued to produce four-cylinder motorcycles based on his designs until America entered World War II in 1941.

1931 HENDERSON MODEL KJ

The Model K was the first new motorcycle from the Henderson Company following its takeover by Schwinn and was designed by Arthur Lemon. It was noticeably updated compared to what had gone before because of the switch to side-valves rather than the inlet-over-exhaust valves. It too was superseded by an even more updated machine in 1929: the Model KJ. It retained the same capacity as the earlier K models but featured a stronger crankshaft assembly. The cycle parts too were redesigned with more modern tanks and a restyled frame being incorporated. Sadly, the motorcycle was short-lived due to the fact that Schwinn withdrew from the motorcycle trade in 1931.

SPECIFICATION
Country of origin: USA
Capacity: 1301cc (79.36cu. in.)
Engine cycle: 4-stroke
Number of cylinders: 4
Top speed: 100mph (161kph)
Power: 40bhp
Transmission: 3-speed
Frame: Steel tubular

BELOW: The side-valve 1931 Model KJ was the last four-cylinder model produced by the Henderson company which had been owned by Ignatz Schwinn since 1917. Schwinn withdrew from motorcycle production later in that same year of 1931.

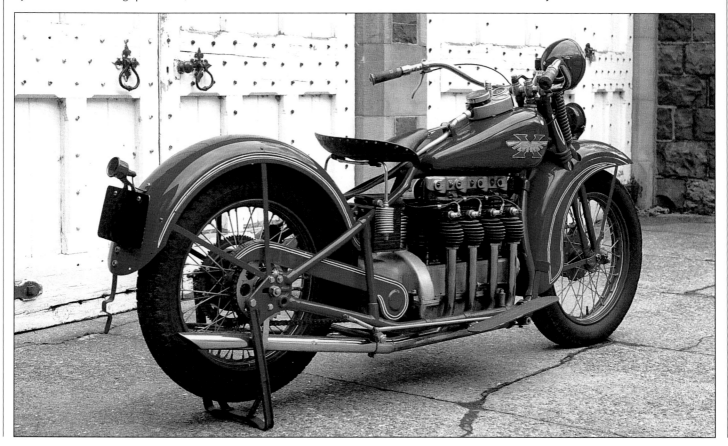

HERCULES

This German company was founded by Carl Marschütz who started in business in 1903, based in Nuremburg, Germany. Over the years Hercules has used a wide range of proprietary engines in their products including Fafnir, JAP, Villiers, Sachs, Ilo, Bark and Moser to name only a few. The company tended to concentrate on smaller capacity motorcycles and during the 1930s built machines with various displacements of less than 500cc (30.50cu. in.). Two of these were the JAP side-valve 198cc (12.07cu. in.) of 1930 and the 197cc (12.01cu. in.) Moser overhead-valve of 1931. Following

World War II the concern manufactured two-stroke models of 248cc (15.12cu. in.) and less, including the two-stroke Ilo 198cc (12.07cu. in.) of 1951. The company grew into one of the larger German motorcycle producers and was bought by Fichtel and Sachs. In 1966 it merged with the Zweirad Union which included DKW, Victoria and Express. In the 1970s the company pro-duced the W2000 rotary which was marketed in some countries as the DKW.

BELOW: As part of the Zweirad Union in Germany, Hercules produced the W2000 in the 1970s, alongside its more traditional motorcycles. The W2000 was a rotary-engined machine which was sold in some export markets as the DKW.

1981 HERCULES ULTRA 80

The Hercules company kept more traditional two-strokes in production alongside the rotary and one was the Ultra 80 which was intended as a simple means of transport. A small machine, it had larger sports bike styling in keeping with the times.

SPECIFICATION
Country of origin: GERMANY
Capacity: 79.8cc (4.86cu. in.)
Engine cycle: 2-stroke
Number of cylinders: 1
Top speed: 52mph (85kph)
Power: 8.5bhp @ 6000rpm
Transmission: 5-speed
Frame: Single tubular cradle

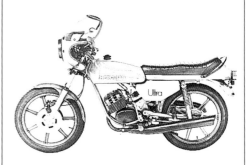

ABOVE: The Hercules Ultra of 1981 was a small capacity motorcycle with sports bike styling typical of its time.

1991 HERCULES KX 5 50

Another small capacity machine that Hercules produced to serve as basic transportation is the KX 5 50 of 1991. Like the Ultra 80 it has sports bike styling in the form of a nose fairing and alloy wheels.

SPECIFICATION
Country of origin: GERMANY
Capacity: 49cc (2.98cu. in.)
Engine cycle: 2-stroke
Number of cylinders: 1
Top speed: 31mph (50kph)
Power: 4bhp @ 5000rpm
Transmission: 5-speed
Frame: Duplex tubular cradle

HESKETH

This company was a relatively short-lived attempt to revitalize the much lamented British motorcycle industry by an English

LEFT: The Hesketh Vampire was a fully-faired touring version of the Weslake V-twin-engined motorcycle produced by the English Lord Hesketh. The company experienced numerous problems and few bikes were built.

lord of the same name. The motorcycle was launched in 1980 but production did not start until 1982 as a result of a number of engineering defects that had to be rectified in order to make the machine viable. The manufacturer, Hesketh Motorcycles Ltd., went into receivership after a total of less than 150 motorcycles were manufactured. Other companies including Hesleydon Ltd. and Mick Broom Engineering have sought to continue production.

1982 HESKETH V1000

The engine for the V1000 was developed by Weslake for the new company. It was a 90° V-twin which featured double overhead camshafts. The remainder of the motorcycle used a number of proven components bought in from other sources, including Marzocchi forks, and the whole was assembled around a duplex tubular cradle frame. A later model based around the same basic components was a tourer, the fully-faired Hesketh Vampire, a motorcycle which was launched in June 1983.

SPECIFICATION
Country of origin: GREAT BRITAIN
Capacity: 992cc (60.51cu. in.)
Engine cycle: 4-stroke
Number of cylinders: 2
Top speed: 130.5mph (210kph)
Power: 86bhp @ 6500rpm
Transmission: 5-speed
Frame: Duplex tubular cradle

BELOW: The Hesketh V1000 of 1981 featured a 90° V-twin engine with double overhead camshafts and was intended to be a new British sports bike. However, it ran into production difficulties which delayed its introduction until 1982.

HILDEBRAND AND WOLFMÜLLER

This company, who operated in Germany between 1894 and 1897, have the distinction of offering the world's first commercially-produced motorcycle. It was also the first machine to be called a motorcycle, or Motorrad as it was termed in German. Heinrich Hildebrand and Alois Wolfmüller designed and built it with a mechanic named Hans Geisenhof. The company had established a number of branch factories but only remained in business for four years. As a result there is some uncertainty as to how many motorcycles the company actually built.

1894 HILDEBRAND AND WOLFMÜLLER

This machine comprised a steel frame of an unorthodox design – not that what was orthodox had been established by this time – and a flat four-stroke, water-cooled, twin-cylinder engine. It had a capacity of 1488cc (90.76cu. in.) through a bore and stroke of

BELOW: The unorthodox appearance of the 1894 Hildebrand and Wolfmüller motorcycle shows that steam engine technology was used in its design and manufacture.

90 and 117mm. The connecting rods followed steam engine practice in that they were connected directly to the rear wheel which doubled as the flywheel, and also through timing devices which opened the valves at the right moment. Tensioned rubber bands assisted the rods to return while the rear fender incorporated the radiator to cool the engine. The engine also featured total loss lubrication and glow combustion

SPECIFICATION
Country of origin: GERMANY
Capacity: 1488cc (90.77cu. in.)
Engine cycle: 4-stroke
Number of cylinders: 2
Top speed: 28mph (45kph)
Power: 2.5hp @ 240rpm
Transmission: Single-speed
Frame: Steel tube

HODAKA

In 1961 PABATCO (Pacific Basin Trading Company) started the import of motorcycles from the Yamaguchi concern of Japan. This Japanese company went out of business in 1963 but the Hodaka Industrial Company of Nagaya, who had formerly been making the engines and transmissions for PABATCO, decided to build entire bikes and started with a trail bike design. This was designed to suit the American concept of dirt-riding, i.e. it had to be dual purpose, capable of covering difficult terrain and also be suitable for road use. Their first road model was the Ace 90 and there then followed a series of rather amusingly-named machines: the Super Rat, the Wombat, the Road Toad and the Dirt Squirt. The company produced these machines until 1978, by which time it could no longer compete with the similar machines being manufactured by the big Japanese companies.

1977 HODAKA 250/SL

Hodaka dirt bikes became popular in the USA and were regularly seen in American motocross and enduro competitions. The engines featured an iron cylinder bore that permitted reboring with an aluminum outer set of fins to permit sufficient cooling. The 250/SL was the larger of the 1977 range which also included the 125/SL and the 100 Road Toad.

SPECIFICATION
Country of origin: JAPAN
Capacity: 246cc (15cu. in.)
Engine cycle: 2-stroke
Number of cylinders: 1
Top speed: n/a
Power: n/a
Transmission: 5-speed
Frame: Tubular steel cradle

HONDA

Honda is the world's biggest motorcycle manufacturer yet the company started after World War II from the humblest of origins. In 1948 Sochiro Honda began by fitting army surplus engines into bicycles. The first Honda engine was a 50cc (3.05cu. in.) two-stroke engine and by 1949 the company was in a factory in Hamamatsu, Japan. During the 1950s the company switched to producing four-stroke engines and in 1958 introduced the 50cc (2.90cu. in.) C-100 with a step-through frame. In 1959 the company opened an American office and began to work toward penetration of the American market. The company became involved with motorcycle racing in 1960 and in the next seven years won 137 Grand Prix and 18 manufacturers' trophies. A gap in racing ensued until the mid-1970s when the company returned with endurance racing machines. Within 20 years Honda production had reached 10 million machines including mopeds and motorcycles as well as farm implements and boat engines, cars and trucks. Honda launched the first super-bike in 1969 – the CB750 – which led to a variety of displacement four-cylinder machines including the CB400 – the so called 400/4. Honda led the way again in

1974 with the introduction of the heavy-weight Goldwing.

Honda has factories at Hamamatsu, Suzuka and Kumamoto, Japan, and exports approximately 70 percent of its annual output. Honda also have a number of overseas factories which manufacture a variety of machines, including the Goldwings which are made in the USA.

1963 HONDA CB72

This, the Honda Dream Supersport, was one of Honda's first sports bikes and one of the first to make inroads into export markets. The CB72 was a road model while a street scrambler variant was tagged the CL72. Both motorcycles featured the same overhead camshaft, parallel-twin engine although some of the cycle parts varied to reflect the different styles of machine.

SPECIFICATION
Country of origin: JAPAN
Capacity: 249cc (15.18cu. in.)
Engine cycle: 4-stroke
Number of cylinders: 2
Top speed: 80mph (128kph)
Power: 24bhp @ 9000rpm
Transmission: 4-speed
Frame: Tubular steel cradle

ABOVE: The Honda CB72 of 1963 was one of the first Honda models intended to be sold seriously in the company's new export markets. The four-stroke twin was conventional in its design with drum brakes and a parallel-twin engine but its success soon established Honda as a major manufacturer.

1965 HONDA CB160

Honda followed the success of their early models with new machines such as the CB160. The numerical designation reflected the approximate metric displacement of the parallel-twin engine. The engine itself was advanced, featuring as it did an overhead camshaft and an electric starter which was something of a novelty in the mid-1960s. Honda claimed the frame and brakes were race-bred which made the machine safe for street riding.

SPECIFICATION
Country of origin: JAPAN
Capacity: 161cc (9.82cu. in.)
Engine cycle: 4-stroke
Number of cylinders: 2
Top speed: 85mph (140kph)
Power: n/a
Transmission: 4-speed
Frame: Tubular steel

ABOVE: The CB160 was launched by Honda in 1965 following the success of the larger displacement CB72.

1965 HONDA CB77

The CB77, at a little over 300cc displacement, was the biggest bike in Honda's range on its introduction in 1964 and it was a motorcycle which established their reputation as builders of more than just small capacity mopeds. It was intended for high-speed touring and its tubular frame used the engine as a stressed member.

SPECIFICATION
Country of origin: JAPAN
Capacity: 305cc (18.60cu. in.)
Engine cycle: 4-stroke
Number of cylinders: 2
Top speed: 100mph (161kph)
Power: 27bhp @ 9000rpm
Transmission: 4-speed
Frame: Tubular steel

ABOVE: The CB77 was styled in a similar way to the CB72 but displaced 305cc (18.60cu. in.) and was, in 1965, Honda's largest capacity motorcycle. It featured an unusual frame but was nonetheless a success.

BELOW: The 'Monkey Bike' models include the ST series such as this ST 70 of 1977. Various displacements of the ST were manufactured.

1977 HONDA ST 70 MONKEY BIKE

This was another motorcycle where the numerical suffix approximated to the metric displacement of the machine's engine. The ST 70 Monkey Bike, though, was an unusually small motorcycle that rolled on 10in. (250mm) diameter wheels which made it seem even smaller. It was designed as basic transportation and despite its diminutive size was intended to carry two people.

SPECIFICATION
Country of origin: JAPAN
Capacity: 72cc (4.39cu. in.)
Engine cycle: 4-stroke
Number of cylinders: 1
Top speed: 40mph (64kph)
Power: 6bhp @ 9000rpm
Transmission: 3-speed
Frame: Tubular steel

ABOVE: As well as small capacity motorcycles, Honda built machines that were physically small. Size was kept to a minimum through use of scooter-sized wheels and miniature components. The motorcycles were known as 'Monkey Bikes'. This is a 1961 CZ100.

1977 HONDA CUB 50

This step-through design of motorcycle was the first from Honda, who intended the 50cc (3.05cu. in.) machine as basic transportation. It appeared in 1958 and, slightly upgraded, it was in production nearly 40 years later. Its attributes are economy, cleanliness, quiet running, and cheapness of purchase. The Cub uses a plastic legshield that bolts over the tubular steel spine which is composite with pressed steel sections further back. The main upgrade in the production run has been the shift from an overhead-valve engine to an overhead camshaft type in 1966. A variety of displacements have been offered including 50, 70 and 90cc (3.05, 4.27 and 5.49cu. in.) versions. For a brief period at the end of the 1960s, as the market for trail bikes grew, a trail version was offered that came with knobbly tires and without the plastic legshields. The fuel tank is mounted under the seat, the chain is enclosed, and the front forks are of a leading link design built in pressed steel. It is the biggest selling motorcycle ever and has been exported all over the world. It is estimated that over 21 million Cubs of the various types have been made.

ABOVE: The Honda Cub, a step-through moped, is the world's most mass-produced motorcycle. It was introduced in 1958 and has been in production, with slight upgrades, at various Honda plants ever since.

SPECIFICATION
Country of origin: JAPAN
Capacity: 89.5cc (5.45cu. in.)
Engine cycle: 4-stroke
Number of cylinders: 1
Top speed: 50mph (80kph)
Power: 7.5bhp @ 9500rpm
Transmission: 3-speed
Frame: Pressed steel

1978 HONDA CB750 F2

The original CB750 was known as the F1 and started the superbike boom when it was introduced in 1969. However, it was upstaged by Suzuki's GS750 Four in 1976 so Honda brought out the CB750 F2. The CB750 features an in-line, four-cylinder, overhead camshaft engine mounted transversely in the frame. While the bike is traditionally styled, it was modern for its time with twin disc front brakes and spoked composite construction wheels.

ABOVE: The CB750 was introduced in 1969 but a much updated CB750 is still in production.

RIGHT: The CB750F of 1979 featured slab styling that was a feature of Honda's entire range at that time.

SPECIFICATION
Country of origin: JAPAN
Capacity: 736cc (44.89cu. in.)
Engine cycle: 4-stroke
Number of cylinders: 4
Top speed: 124.6mph (200.5kph)
Power: 70bhp @ 9000rpm
Transmission: 5-speed
Frame: Steel duplex cradle

wanted and progress to larger versions if they so desired. The CG125 has had a lengthy production run and has been only minimally upgraded in more than a decade.

SPECIFICATION
Country of origin: JAPAN
Capacity: 124cc (7.56cu. in.)
Engine cycle: 4-stroke
Number of cylinders: 1
Top speed: 60mph (96.5kph)
Power: 11bhp @ 9000rpm
Transmission: 5-speed
Frame: Single tubular cradle

1982 HONDA CB250NDX-B

The Superdream, as the CB250N was known, evolved out of an earlier series of Honda motorcycles called Dreams. The first one was a 305cc (18.60cu. in.) motorcycle of the early 1960s. This was followed by a series of vertical twins known as Honda Dreams which were part of a comprehensive series of bikes designated CB. CB models ranged from 125cc (7.62cu. in.) to the CB750 and included the CB200 and the CB 400. The Superdream was a restyled bike which had an angular design to make it more modern in appearance. The range was redesigned so that the Superdream looked like a scaled-down version of the CB900, an in-line four-cylinder motorcycle. A larger

variant, the CB400N-B, was also available that displaced 395cc (24.09cu. in.).

SPECIFICATION
Country of origin: JAPAN
Capacity: 249cc (15.18cu. in.)
Engine cycle: 4-stroke
Number of cylinders: 2
Top speed: 85mph (140kph)
Power: 27bhp @ 10,000rpm
Transmission: 6-speed
Frame: Diamond open cradle

ABOVE AND BELOW: The Honda CBX 1000 was introduced in 1978 and was fitted with a transverse six-cylinder engine. Air-cooling of the complex engine was aided by the engine being inclined forward. The four-stroke was one of the superbikes of its day and was capable of almost 140mph (225kph). The CBX was one of only three popular transverse six-cylinder bikes of its time. The others were the Z1300 Kawasaki and the Benelli Sei. More recently, Honda introduced the six-cylinder engine for the Goldwing and Valkrie models.

1982 HONDA CBX-1000

Honda introduced the CBX 1000 in 1978. It was seen by many people as an extravagance because it featured an across-the-frame, six-cylinder engine which had twin overhead camshafts. Honda minimized the width by moving all the ancillary engine parts away from the crankshaft ends – the alternator, for example, was mounted behind the bank of cylinders. The engine itself was inclined forward to assist cooling but it was a complex mechanism and featured 24 valves, i.e. four per cylinder, and six carburetors. The design and, indeed, the overall styling of the machine reflected a number of other motorcycles in Honda's range at that particular time with five-spoke composite wheels and an angular gas tank. In 1981 the motorcycle was redesigned and a fairing was added.

SPECIFICATION
Country of origin: JAPAN
Capacity: 1047cc (63.86cu. in.)
Engine cycle: 4-stroke
Number of cylinders: 6
Top speed: 137mph (220kph)
Power: 100bhp @ 9000rpm
Transmission: 5-speed
Frame: Tubular spine

RIGHT: CB250 is another long running model designation from Honda. The original machines bear little resemblance to this model from the early 1990s.

1986 HONDA CB250

The CB250 is a long-running model designation for what is a medium-weight, mid-size 250cc (15.25cu. in.) Honda twin. The engine has been modified several times and the styling revised in line with Honda's current range. The motorcycle started out on its life with curved tanks which gradually became more angular as the years passed and fashions changed.

SPECIFICATION
Country of origin: JAPAN
Capacity: 249cc (15.18cu. in.)
Engine cycle: 4-stroke
Number of cylinders: 2
Top speed: 102mph (164kph)
Power: 31bhp @ 9500rpm
Transmission: 6-speed
Frame: Tubular single cradle

1991 HONDA XVR750 AFRICA TWIN

Honda based this model on their Paris-Dakar race winner, the NXR. Its styling closely follows those bikes with features such as twin headlamps in a small fairing and an engine guard. The suspension at the rear is monoshock and, at the front, tele-scopic forks. Disc brakes are fitted front and rear. Ground clearance is enhanced through the use of 19 and 17in. wheels (475 and 470mm) front and rear respectively. The engine is a liquid-cooled, three-valves-per-cylinder V-twin and, like the rest of the machine, designed to enable the XVR750 to speed across unsurfaced terrain. Another

BELOW: The Africa Twin was Honda's version of a type of motorcycle that became enormously popular with the success of 'raid' events such as the annual Paris-Dakar marathon. Motorcycles such as this model were a compromise between high performance and off-road ability, hence the appearance of being an oversized trail bike.

SPECIFICATION
Country of origin: JAPAN
Capacity: 748cc (45.62cu. in.)
Engine cycle: 4-stroke
Number of cylinders: 4
Top speed: 137mph (220kph)
Power: 100bhp @ 10,500rpm
Transmission: 6-speed
Frame: Aluminum box section

1991 HONDA CBR 600

This was a middleweight sports bike that was introduced to great acclaim in 1987. It has been upgraded on a number of occasions to keep it at the top of the sales charts, including a complete revision in 1991. Its engine is very typical of 1990s Honda: liquid-cooled, double overhead camshaft, 16-valve, four-stroke, four-cylinder. The CBR 600 sits on 17in. (470mm) diameter wheels and is fully faired.

SPECIFICATION
Country of origin: JAPAN
Capacity: 599cc (36.53cu. in.)
Engine cycle: 4-stroke
Number of cylinders: 4
Top speed: 152mph (245kph)
Power: 100bhp @ 12,000rpm
Transmission: 6-speed
Frame: Twin spar diamond

similar model from Honda was the NX650 Dominator that was powered by a single-cylinder four-stroke engine.

SPECIFICATION
Country of origin: JAPAN
Capacity: 742cc (45.26cu. in.)
Engine cycle: 4-stroke
Number of cylinders: 2
Top speed: 112mph (180kph)
Power: 61.4bhp @ 7500rpm
Transmission: 5-speed
Frame: Steel box section cradle

1991 HONDA NSR 125

This is a small capacity, two-stroke single sports bike aimed at new riders who aspire to larger machines. It is fully faired with race-replica styling and produced by Honda in Italy. Cast alloy wheels, monoshock rear suspension and vented disc brakes emphasize the sports aspect of the bike.

SPECIFICATION
Country of origin: ITALY
Capacity: 124.8cc (7.61cu. in.)
Engine cycle: 2-stroke
Number of cylinders: 1
Top speed: 72mph (117kph)
Power: 12bhp @ 7500rpm
Transmission: 6-speed
Frame: Aluminum cast diamond

RIGHT: The NSR 125 is a small capacity motorcycle with sports bike styling aimed at younger riders who aspire to bigger motorcycles. It has many of the features of the larger machines such as an aluminum frame.

1991 HONDA VFR750F (V4)

The VFR750F is one of the 1990s generation of roadgoing sports bikes featuring an angular full fairing and an aluminum frame. Prominent among the other high-tech features included in the design are the TRAC anti-dive system in the air-assisted forks, the monoshock rear suspension system, the cast alloy wheels and the huge disc brakes at both front and rear.

ABOVE: The Honda VFR750R is a four-stroke,
liquid-cooled, V-four motorcycle with a 748cc
(45.62cu. in.) displacement through bore and
stroke of 70 and 48.6mm. It is capable of a top
speed of 137mph (220kph).

HONG DU

Hong Du are a Chinese motorcycle manufacturer who started in business during the mid-1960s. The company produces small capacity, two-stroke machines in association with the larger Japanese manufacturers such as Yamaha. Their motorcycles are aimed at providing basic transportation within the People's Republic of China although some small numbers have been exported.

HOREX

Horex was a German company based at Bad Homburg. It was founded in the early 1920s by the Kleeman family who were in other production businesses. Their motorcycles used Columbus proprietary engines. Their first machine was a 246cc (15cu. in.) overhead-valve single. A range of larger displacement motorcycles followed and in the 1930s their designer, Hermann Reeb, designed a pair of vertical twin engines that had chain-driven camshafts. The engines were considered technologically advanced and had some racing success, especially with Karl Braun on a supercharged version.

In the years which followed World War II Horex went back to motorcycle production and the company introduced a successful 349cc (21.28cu. in.) machine called the Horex Regina. Later this was superseded by the Horex Imperator. Production by the company stopped in 1958. Later the name reappeared, with the company owned by another: Zweirad Röth GmbH & Company of Odenwald, Germany.

1986 HOREX HRD 600

The idea of Horex using proprietary engines is nothing new and in 1986 they produced the HRD 600 and 500 based around the successful Rotax engine in two capacities of 494 and 562cc (30.13 and 34.28cu. in.). The HRD 600 was a modern bike for its day using monoshock rear suspension and air-assisted front forks.

SPECIFICATION
Country of origin: GERMANY
Capacity: 562cc (34.28cu. in.)
Engine cycle: 4-stroke
Number of cylinders: 1
Top speed: n/a
Power: n/a
Transmission: 5-speed
Frame: Chrome molybdenum tubular

HRD

This company bears the initials of its founder, Howard R. Davies, who had been apprenticed to AJS and had also raced

ABOVE: **Many of Howard R. Davies' motorcycles, known as HRD, used single-cylinder proprietary engines of varying capacities manufactured by JAP.**

Sunbeam motorcycles prior to World War I. In 1921 he won the Isle of Man Senior TT aboard a 350cc (21.35cu.in.) AJS and went on to found HRD in 1924. HRD motorcycles were designed by E. J. Massey and used JAP single-cylinder engines of 344 and 490cc (20.98 and 29.89cu. in. respectively). The company ceased production when they went into liquidation in 1927 after manufacturing a 597cc (36.41cu. in.) machine. Philip Vincent purchased the rights from OK Supreme who had purchased them when HRD closed down. The company then became Vincent–HRD.

HUMBER

Thomas Humber founded a company in 1870 to build pedal bicycles. It was established at Beeston, near Nottingham, England. By 1898 he was experimenting with motorized bicycles and tricycles. In 1902 the company produced a single-cylinder, 344cc (21cu. in.) motorcycle. The engine was designed by a Jonah Phelon, who would later found P & M in West Yorkshire, England. This motorcycle featured an engine with an inclined cylinder that took the place of the front downtubes. By 1911 Humber motorcycles were proving so successful that one ridden by P. J. Evans won the Junior TT on the Isle of Man. In the years after World War I a range of Humbers existed including flat-twins and

BELOW: **Humber were typical of many English manufacturers in that their origins were in bicycle manufacture and they also made cars. This 2.75hp model was made in 1925.**

both overhead and side-valve singles. The company stopped motorcycle manufacture in 1930 and from that point onward concentrated on car production.

HUSQVARNA

Originally, Husqvarna was a Swedish arms manufacturer that also made bicycles. In 1903 they started making motorcycles by fitting NSU and FN engines into machines of their own design. The company then progressed in 1920 to an engine of its own design, a 550cc (33.55cu. in.) side-valve V-twin. They continued to use proprietary engines including those from JAP. They increased the displacement of their own engines to as much as 1000cc (61cu. in.). In the mid-1930s Husqvarna made its first two-stroke – a 98cc (5.98cu. in.) motorcycle with a two-speed gearbox. It sold well and after World War II was over the company concentrated on such small machines. However, as popular demand for off-road motorcycles grew Husqvarna moved successfully into this field. It used Albin engines in its four-strokes and its own engines in the

BELOW: Husqvarna manufactured V-twin motorcycles for racing between 1932 and 1935 – a 1934 model is shown. The V-twins won a succession of Swedish Grands Prix and later the company also found a strong measure of success in motocross competition.

ABOVE: The Model 30 TV was one of a range of motorcycles produced by Husqvarna that used proprietary engines in both side- and overhead-valve configuration and with both two- and four-stroke engines.

two-strokes. Riders on Husqvarna machines had enormous success in motocross competition. The company developed its bikes to include automatic transmissions and liquid-cooling. In 1986 the company was acquired by Cagiva, the successful Italian group of companies, and production was transferred there. A number of former Husqvarna employees produced off-road competition bikes under the Husaberg brandname which was established in 1988.

BELOW: Husqvarna had much competitive success after World War II when it manufactured a range of two-stroke motorcycles for off-road competition. This enduro bike dates from the 1980s.

1986 HUSQVARNA MC250MP

This two-stroke motorcycle was built by Husqvarna for the Swedish armed forces and is designed to fulfil a multi-purpose role for the military. The machine's dirt bike heritage is clearly obvious and the bike can be fitted with skis to enable it to be used on NATO's northern flank in wintertime.

SPECIFICATION
Country of origin: SWEDEN
Capacity: 250cc (15.25cu. in.)
Engine cycle: 2-stroke
Number of cylinders: 1
Top speed: 68mph (110kph)
Power: 20bhp
Transmission: 4-speed
Frame: Tubular cradle

1986 HUSQVARNA 125 WRK

The smaller capacity market for off-road bikes was not ignored by Husqvarna who manufactured the 125WRK with a 124cc (7.56cu. in.) displacement engine. The motorcycle was designed for motocross and featured high ground clearance and long travel suspension through use of an alloy swinging arm and Ohlins monoshock and telescopic leading axle forks. Tall tires and

ABOVE: The WRK 125 Cross was a small capacity motorcycle designed for off-road competition and used as many up-to-date components as the larger capacity models.

wheels which were 21in. and 18in. (530 and 450mm) front and rear respectively also increased the bike's ground clearance. Mudguards and fuel tank were plastic.

SPECIFICATION
Country of origin: SWEDEN
Capacity: 124cc (7.56cu. in.)
Engine cycle: 2-stroke
Number of cylinders: 1
Top speed: n/a
Power: n/a
Transmission: 6-speed
Frame: Chrome molybdenum cradle

1991 HUSQVARNA 610 TE

This motorcycle was made as one of the generation of four-stroke singles used in motocross events. The 610 TE is a modern, technologically advanced machine featuring as it does a liquid-cooled, double overhead camshaft, four-valve engine. The remainder of the machine is as modern as the engine and consists of a tube steel frame fitted with an alloy swingarm and monoshock suspension assembly. The forks are of the upside-down type and disc brakes are fitted front and rear.

SPECIFICATION
Country of origin: ITALY
Capacity: 577cc (35.19cu. in.)
Engine cycle: 4-stroke
Number of cylinders: 1
Top speed: n/a
Power: n/a
Transmission: 6-speed
Frame: Tubular cradle

BELOW: The 610 TE Husqvarna of 1991 is a technologically advanced motocrosser with a liquid-cooled, double overhead camshaft, four-valve, single-cylinder engine in a modern tubular cradle frame with an alloy swing arm and monoshock suspension unit.

INDIAN

This American company was established in Springfield, Massachusetts in 1901 by George Hendee and Oscar Hedstrom who had met through a mutual interest in cycle racing. Their first machine was introduced in 1901 and featured the inlet-over-exhaust configuration and total loss lubrication that was common to the majority of the early machines. Indian's first machine utilized a bicycle-style diamond frame with a single-cylinder engine installed within. The red that found fame as the distinctive Indian Red was introduced in 1904 and twin-cylinder motorcycles were introduced in 1907, through use of a V-twin designed by Oscar Hedstrom. By 1909 Indian had produced a loop frame as motorcycles and pedal cycles diverged. In 1911 Indian had a remarkable success in the Isle of Man TT races when their machines took first, second and third places in that year's Senior race. Indian survived World War I and they produced a huge number of machines for the US Army. In these years Hendee and Hedstrom resigned from the company they had founded having become dissatisfied with the way successive boards of directors were running it. A new generation of machines followed from employed engineers such as Charles Gustafson who designed the Powerplus. This machine stayed in production, albeit sequentially upgraded, until 1924. The next generation of motorcycles

ABOVE: The first Indian motorcycles were painted blue rather than the red for which Indian later became famous. This particular machine was made in 1902 and features an inlet-over-exhaust engine of 225cc (13.72cu. in.) displacement that produced 1.75bhp.

from Indian were designed by Charles Franklin, of which the most notable were the Scout and Chief. These models were introduced in 1920 and 1922 respectively to immediate acclaim. The Chief was initially billed as a Big Scout and considered suitable for sidecar work. Like the smaller

displacement Scout, the Chief endured for many years, being gradually upgraded and developed – indeed it was the last American-made Indian to be produced. During World War II Indian supplied militarized motorcycles to the US Army. These machines were essentially military variants of the Scout models. During this period Indian also built an experimental machine, the 841, for the US Army, which featured shaft drive, a transverse V-twin engine and plunger rear suspension. While it was a successful motorcycle – as was its competitor, Harley–Davidson's flat-twin XA – the army did not order either machine in any large quantity.

1904 INDIAN SINGLE

Indian started out as a brand of motorcycle which was manufactured by the Hendee Manufacturing Company in 1901. The earliest bikes, such as the 1904 Indian Single, show their bicycle origins with unsprung forks and a diamond bicycle-type frame.

SPECIFICATION
Country of origin: USA
Capacity: 288.2cc (17.59cu. in.)
Engine cycle: 4-stroke
Number of cylinders: 1
Top speed: 30mph (50kph)
Power: n/a
Transmission: Single-speed
Frame: Diamond

LEFT: This 1904 single-cylinder Indian motorcycle was the 667th ever produced by the company that was founded by Hendee and Hedstrom in 1901. The bicycle origins of the machine are clearly evident.

1909 INDIAN SINGLE

Up until 1909 Indian's motorcycles showed their bicycle heritage through the use of a bicycle-style diamond frame. However, Indian brought out their first loop frame as an option in 1909 and it soon replaced their bicycle-style frame permanently. The loop frame gave a lower center of gravity and was more attractive visually.

BELOW: In 1909 Indian moved away from the bicycle-type diamond frame to the loop frame seen here. The single-cylinder four-stroke engine gave the machine a top speed of 36mph (58kph).

SPECIFICATION
Country of origin: USA
Capacity: 442cc (26.96cu. in.)
Engine cycle: 4-stroke
Number of cylinders: 1
Top speed: 36mph (58kph)
Power: n/a
Transmission: Single-speed
Frame: Loop

1916 INDIAN POWERPLUS

Side-valves had caught on in a big way in European motorcycle design and manufacture but only Reading Standard had really taken and developed the idea in the USA.

ABOVE: The Powerplus of 1920 was powered by a side-valve V-twin engine of 997.6cc (60.88cu. in.) and was capable of 62mph (100kph). The transmission was three-speed.

The company explored the possibilities of side-valves under the direction of Charles Gustafson. However, he soon left Reading Standard and moved on to Indian where he designed a side-valve engine for the new Powerplus model. Gustafson's new design was much more powerful than Indian's earlier ones. and it became the basis of the company's continued success until 1953 when the company closed down.

SPECIFICATION
Country of origin: USA
Capacity: 997.6cc (60.88cu. in.)
Engine cycle: 4-stroke
Number of cylinders: 2
Top speed: 60mph (96.5kph)
Power: 18bhp
Transmission: 3-speed
Frame: Steel loop

1917 INDIAN MODEL O

Indian were not alone in believing that there was a market for lightweight motorcycles of small displacement. In their range for several years were lightweight machines including the Model O, which was in production for two years between 1917 and

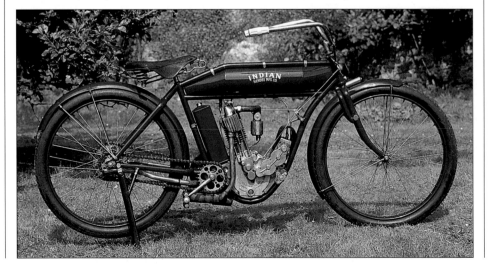

1919. This diminutive bike featured a side-valve, horizontally-opposed twin engine.

SPECIFICATION
Country of origin: USA
Capacity: 257.3cc (15.67cu. in.)
Engine cycle: 4-stroke
Number of cylinders: 2
Top speed: 45mph (72kph)
Power: 4bhp
Transmission: 3-speed
Frame: Steel loop

1928 INDIAN FOUR

The Indian Four was a magnificent motorcycle and its purchase price reflected this. The Four came about as Indian acquired the rights and tooling to the Ace motorcycle when Ace ceased trading. The first Fours were referred to as Indian Aces although this tag was dropped as Indian's engineers refined the design.

SPECIFICATION
Country of origin: USA
Capacity: 1265cc (77.16cu. in.)
Engine cycle: 4-stroke
Number of cylinders: 4
Top speed: 80mph (128kph)
Power: 30bhp
Transmission: 3-speed
Frame: Steel loop

ABOVE: The first Indian Fours were referred to as Indian Aces because Indian had in 1927 acquired the assets of The Ace Motorcycle Company and commenced production of them.

1928 INDIAN 101 SCOUT

Indian Scout reliability became noteworthy and gave rise to the advertising slogan, 'You can't wear out an Indian Scout'. The Scout went through a number of incarnations, all of which brought racing success to the company in its campaigns against arch-rival bike builders Harley–Davidson. Many riders felt that the 101 Scout was the best motorcycle Indian ever manufactured due to it being endowed with handling characteristics that became legendary.

SPECIFICATION
Country of origin: USA
Capacity: 737cc (44.95cu. in.)
Engine cycle: 4-stroke
Number of cylinders: 2
Top speed: 75mph (120kph)
Power: 18bhp
Transmission: 3-speed
Frame: Steel loop

BELOW: The Indian 101 Scout, such as this 1928 model, was noted for its good handling which brought it numerous race victories.

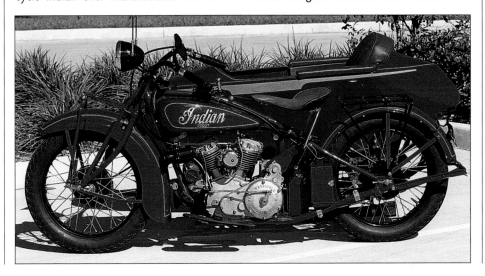

1935 INDIAN SPORT SCOUT

The Sport Scout was Indian's sports bike which was brought out in response to the increasing customer and dealer demand for a new model to replace the discontinued 101 Scout. It was introduced in 1934 and had a European look about it. This was as a result of its girder forks and the distinctive design of its frame that was termed a 'keystone' by its makers because of the way it relied on the engine to be part of it. The

Sport Scout became the basis of successful Indian racers of the era.

SPECIFICATION
Country of origin: USA
Capacity: 744cc (45.38cu. in.)
Engine cycle: 4-stroke
Number of cylinders: 2
Top speed: 80mph (128kph)
Power: 25bhp
Transmission: 3-speed
Frame: Keystone steel

ABOVE: The Sport Scout debuted in 1934 after the discontinuation of the 101 Scout. It was popular as a race bike because of its keystone frame. The roadgoing versions were equipped with large valanced fenders immediately before World War II.

1942 INDIAN 841

This was an experimental Indian motorcycle built by the company at the request of the US Government in the same way as Harley–Davidson built the XA. The 841 featured a transverse V-twin engine, shaft drive, girder forks and plunger rear suspension. These features made it a technologically advanced motorcycle for its time but the army never ordered them in any really significant numbers.

SPECIFICATION
Country of origin: USA
Capacity: 750cc (45.75cu. in.)
Engine cycle: 4-stroke
Number of cylinders: 2
Top speed: n/a
Power: n/a
Transmission: 3-speed
Frame: Duplex cradle

LEFT: The Indian transverse V-twin, which was known as the Model 841, was only ever made in very limited numbers because of its experimental nature. In all, less than 1000 motorcycles were manufactured.

ABOVE: In the years following World War II Indian resumed production of the Indian Chief. The last of its Chiefs, made between 1947 and 1953, were luxury motorcycles with girder forks and plunger rear suspension, as seen on this 1948 model.

1948 INDIAN CHIEF

Although the Indian Chief was launched in the 1920s the name brings to mind the opulent cruisers that were the last of the line. What made the Chief so distinctive were the fully valanced fenders that partially hid the wheels. This style of fender was introduced immediately before World War II and reintroduced after it. The Indian Chief itself was reintroduced postwar with girder forks and plunger rear suspension.

SPECIFICATION
Country of origin: USA
Capacity: 1206cc (73.56cu. in.)
Engine cycle: 4-stroke
Number of cylinders: 2
Top speed: 85mph (140kph)
Power: 40bhp
Transmission: 3-speed
Frame: Duplex tube

1951 INDIAN BRAVE

Indian were eventually acquired by a British company, Brockhouse Engineering, and in the immediate postwar era it was felt that there was a need for a lightweight motorcycle of a European type. Incidentally, Indian had experimented with lightweight machines for most of their 52 year history, never particularly successfully, and the Brockhouse Brave was no exception. It was made in England, shipped to America and sold bearing the famous Indian logos.

SPECIFICATION
Country of origin: GREAT BRITAIN
Capacity: 248cc (15.12cu. in.)
Engine cycle: 2-stroke
Number of cylinders: 1
Top speed: 58mph (93kph)
Power: n/a
Transmission: 3-speed
Frame: Steel loop

ITALJET

This Italian company dates from the mid-1960s and produces a number of two-stroke machines for off-road use. The company has used proprietary engines from companies such as Minarelli, CZ, MZ, Yamaha and Triumph. The 646cc (39.40cu. in.) engines from this latter company were fitted into Italjet's own frame to give a large capacity off-road machine.

RIGHT: The Italjet Formula 125 of 1996 is an unusual machine that incorporates modern scooter styling with high technology, including a twin-cylinder 125cc (7.62cu. in.) engine.

1982 ITALJET 350T

Typical of early 1980s specialist trials bikes is the Italjet 350T. The designation of this motorcycle approximates to its metric displacement and other machines were similarly designated. These include the 50T, 100T and 250T. Like specialist trials bikes, the 350T features high ground clearance, light weight and low gearing.

SPECIFICATION
Country of origin: ITALY
Capacity: 326cc (19.88cu. in.)
Engine cycle: 2-stroke
Number of cylinders: 1
Top speed: n/a
Power: n/a
Transmission: 6-speed
Frame: Single tubular cradle

ABOVE: The Indian Chief of 1947 was a truly
opulent motorcycle powered by a large
capacity side-valve engine. It featured hugely
valanced fenders fitted to girder forks and a
plunger frame, front and rear respectively.

J

JAMES

This company had its roots in the closing decades of the 19th century and manufactured its first motorcycle in 1902. The company had an innovative style and used engines from FN as well as those of their own design. The factory burned down in 1920 which was something of a setback and production was badly affected for two years. Following this, the company produced singles and V-twins in relatively small numbers. For the 1930s their range of machines included both James- and Villiers-powered motorcycles of small displacement because the company had perceived which way their market was moving. These bikes were conventional with rigid frames, girder forks and were typical of utility machines of the period. A Rudge 499cc (30.43cu. in.) engined machine was offered briefly in 1931. The range stayed as utility bikes from then on, especially after their final V-twin was dropped in 1936. By 1937 all their power units were coming from Villiers.

World War II saw the James factory once again almost destroyed in the blitz but by 1943 the ML – Military Lightweight – was rolling off the production line for use by airborne soldiers. The ML, of which 6000 were manufactured, had a 122cc (7.44cu. in.) engine and became the basis of a civilian model in the postwar years. Later a new range of machines was introduced that included the Captain, Commodore, Comet,

ABOVE: James built orthodox Villiers-engined bikes and unorthodox Renouf-designed ones. This 1913 model is one of the former.

Colonel and Cadet. In the mid-1950s AMC acquired the company and more of that concern's engines appeared in the James range. Production stopped in 1964.

1959 JAMES FLYING CADET

Towards the end of the 1950s James, having been acquired by AMC who made Matchless and AJS machines, made use of this connection by fitting the AMC engine into the model which they had designated as the Flying Cadet. It was a mixture of Cadet cycle parts with the AMC two-stroke engine fitted.

SPECIFICATION
Country of origin: GREAT BRITAIN
Capacity: 149cc (9.08cu. in.)
Engine cycle: 2-stroke
Number of cylinders: 1
Top speed: n/a
Power: n/a
Transmission: 3-speed
Frame: Steel cradle

ABOVE: The 1959 Flying Cadet was a combination of components from both James and AMC who had acquired the company.

JAP

The initials JAP stand for J. A. Prestwich and was the name of a Tottenham, London, England, based company that is primarily famous for supplying engines to other manufacturers. The company was founded in 1904 and manufactured complete motorcycles until 1908 when a decision was made to only build engines for supply to others. In the postwar years the engine factory was taken over by Villiers who were also engaged in supplying their engines to other manufacturers. Under the wing of Villiers, JAP also went on to supply engines to a number of British marques such as Brough and Matchless.

BELOW: Martin–JAP were an English company in business between 1929 and 1957 who specialized in speedway machines. They used proprietary engines supplied by JAP, who also supplied engines to numerous makers from 1908 onward. This JAP-engined Martin–JAP is a 1933 speedway competition motorcycle.

JAWA

This is another motorcycle manufacturer with its origins in the manufacture of arms. Jawa, a Czech company, were in the closing year of the 1920s when they obtained a license to build German-designed Wanderer motorcycles and so started production of a 498cc (30.37cu. in.) overhead-valve, single-cylinder, unit construction motorcycle that also featured a pressed steel frame and forks. The name Jawa comes from the first two letters of Janecek, the proprietor's surname, and of Wanderer, the model name. During the 1930s the company, in association with an English designer, George-William Patchett, produced some successful racing bikes and a 173cc (10.55cu. in.) two-stroke which became popular and established Jawa's reputation. Jawa's production was interrupted by World War II yet after

the war ended they produced some quite sophisticated two-strokes of 248 and 346cc (15.10 and 21.12cu. in.) displacement. These machines also featured automatic clutches, telescopic forks and plunger rear suspension. The company was nationalized after World War II and has produced two-stroke machines ever since including successful speedway competition bikes. There is close co-operation with CZ, which is also a nationalized Czech motorcycle producer, following a 1949 merger.

1978 JAWA 634/6

The 634/6 is utterly typical of Jawa's utility ride-to-work motorcycles which were built as basic transportation with little regard being given to modern styling. This meant that although the machines were exported far beyond Czechoslovakia they were not

ABOVE: **The two stroke 634/6 350 of 1978 was unremarkable but fulfilled Jawa's intentions of providing basic transportation.**

always popular in more fashion-conscious markets.

SPECIFICATION
Country of origin: CZECHOSLOVAKIA
Capacity: 343.47cc (20.95cu. in.)
Engine cycle: 2-stroke
Number of cylinders: 2
Top speed: 84mph (135kph)
Power: 28bhp @ 5000rpm
Transmission: 4-speed
Frame: Steel tube

JIALING

Jialing are a Chinese operation who manufacture Honda XL 125S motorcycles under license. The diminutive trail-type four-stroke motorcycles have a displacement of 125cc (7.62cu. in.) and have 12 volt electrics. Another Chinese company that produces Hondas under license is Xingfu. This concern also builds Honda-designed 125 and 250cc (7.62 and 15.25cu. in.) two-stroke single models.

ABOVE: **Jawa took over the Eso company and have been a successful speedway bike maker ever since. They have exported their machines around the world since the 1950s.**

BELOW: **A double overhead camshaft, twin-cylinder, 498cc (30.37cu. in.) works racer of 1955, manufactured by the Jawa company for road racing.**

BELOW: **Jialing manufacture these small-capacity motorcycles in large numbers under license from Honda. Most go to satisfy the demand for basic transport in China.**

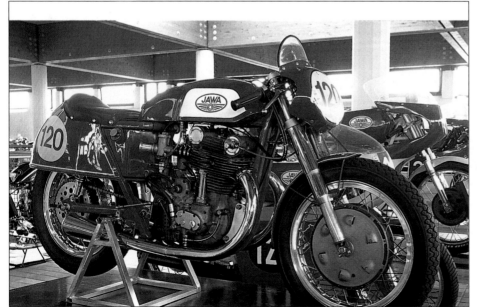

K

KAWASAKI

The company that manufactures Kawasaki motorcycles is part of a industrial conglomerate, called Kawasaki Heavy Industries, that started out building ships. It has also made helicopters, trains, aeroplanes and industrial equipment. Its start in motorcycle manufacture came in the years after World War II when it began supplying small capacity two-stroke engines to a subsidiary known as Meihatsu. During the early 1960s Kawasaki acquired Meguro, a Japanese motorcycle manufacturer that had been in the business since 1937. It added this company's British-styled four-stroke vertical-twins of 624cc (38.06cu. in.) displacement to its range of products. By the mid-1960s Kawasaki was exporting its products to the USA and at the end of the decade intro-

ABOVE: Kawasaki, along with the other 'Big Four' Japanese manufacturers, introduced cruiser-styled motorcycles aimed at Harley–Davidson's share of this market.

BELOW: The four-cylinder, air-cooled Kawasaki made its mark in racing, ensuring popularity with sports riders and specialist companies such as Harris, who built this Z900 in 1975.

duced the models that guaranteed its reputation as producers of sports bikes: its two-stroke triples. These were available as 250, 500 and 750cc (15.25, 30.50 and 45.75cu. in.) models. The further development of these machines was stopped when US emissions regulations made their production less than viable. From this Kawasaki moved on to four-strokes. The Z1 of the early 1970s took the motorcycling world by storm. Just over a decade later the company did it again with the GPZ900 and yet again in 1990 with the ZZR-1100 which was the fastest superbike of its time.

The Kawasaki factory is situated in Kobe, near Akashi, in Japan, where in excess of 250,000 motorcycles are built each year, of which a huge percentage are exported. A US assembly plant was built in Lincoln, Nebraska, in 1975. Unlike other Japanese manufacturers, Kawasaki does not place much emphasis on the utility market and relies on race wins in both circuit racing and motocross to promote its products.

1969 KAWASAKI H1 MACH III

The two-stroke triples from Kawasaki were the machines that established the company's reputation in the motorcycling world. The 250cc (15.25cu. in.) version was launched in 1969 and became renowned for its speed and its handling. The machine was fast for its day and yet could be a difficult bike to ride. The H1 sold well in the USA and was later followed by larger displacement versions.

SPECIFICATION
Country of origin: JAPAN
Capacity: 498cc (30.37cu. in.)
Engine cycle: 2-stroke
Number of cylinders: 3
Top speed: 120mph (193kph)
Power: 60bhp @ 8000rpm
Transmission: 4-speed
Frame: Tubular steel

1972 KAWASAKI 750H2B

The 750H2B was a revelation when it was announced after the H1 and a 500cc (30.50cu. in.) version. The 500 was already renowned for exhilarating performance so that the 748cc (45.60cu. in.) was likely to be unreal. It was the fastest triple-cylinder engine from Japan fitted into a lightweight frame. US emission laws and alarming fuel consumption figures of around 20mpg (32kpg) meant that the 750 triple was discontinued by 1976.

SPECIFICATION
Country of origin: JAPAN
Capacity: 748cc (45.62cu. in.)
Engine cycle: 2-stroke
Number of cylinders: 3
Top speed: 117mph (188kph)
Power: 74bhp @ 6800rpm
Transmission: 5-speed
Frame: Duplex loop cradle

ABOVE: The H2B was a 750cc (45.75cu. in.) displacement triple and the largest in a range that also included 250 and 500cc (15.25 and 30.50cu. in.) machines. All were very fast.

1973 KAWASAKI Z1

Kawasaki introduced the Z1 in 1972 in response to Honda's 750/4 of three years earlier. It was sensational at the time, going further than the Honda in terms of technical engineering. It too was fitted with a transversely-mounted, in-line four-cylinder engine but it also featured double overhead camshafts, something which Kawasaki were keen to trumpet on both the crankcases and sidepanels because the Honda was of a single overhead camshaft design. Liquid-cooling and alloy wheels were very much in the future at the time of the Z1's introduction so the machine was something of a combination of old and new – new in the modernity of the engineering and traditional in terms of the distinctly separate tank and seat units, and wire spoke wheels.

SPECIFICATION
Country of origin: JAPAN
Capacity: 903cc (55.08cu. in.)
Engine cycle: 4-stroke
Number of cylinders: 4
Top speed: 130mph (209kph)
Power: 82bhp @ 8500rpm
Transmission: 5-speed
Frame: Tubular cradle

1977 KAWASAKI Z650B1

The Z650 was Kawasaki's middleweight four-stroke sports bike of the 1970s and it was one which gained the company much acclaim. It became a classic and the Zephyr range of motorcycles were later styled on it and on its larger version, the Z1000A1. Since the Z1, Kawasaki have always had four-stroke, four-cylinder motorcycles in their range although subsequently they switched to liquid-rather than air-cooling.

SPECIFICATION
Country of origin: JAPAN
Capacity: 652cc (39.77cu. in.)
Engine cycle: 4-stroke
Numbers of cylinders: 4
Top speed: 121mph (195kph)
Power: 64bhp @ 8500rpm
Transmission: 5-speed
Frame: Tubular

BELOW: The Z650 was a four-stroke, four-cylinder motorcycle and Kawasaki's answer to Honda's popular 750/4. Four-strokes conform more easily to emission laws and this ensured their popularity.

1979 KAWASAKI Z1300

The Z1300 was not the first across-the-frame six-cylinder motorcycle but it did last longer than the others of the same era, namely the Benelli Sei and the Honda CBX 1000, and it also pioneered certain features on motorcycles. The massive Z1300 was amongst the first liquid-cooled motorcycles of the superbike era. On a later series of Z1300s fuel delivery was by fuel injection which brought increased fuel economy and smoothness of throttle. The power delivery is akin to a luxury car. It is not all instant acceleration but a smooth delivery up to its top speed. The machine is big and weighs around 700lb (317.5kg) but despite this is a conventional motorcycle in appearance.

SPECIFICATION
Country of origin: JAPAN
Capacity: 1286cc (78.44cu. in.)
Engine cycle: 4-stroke
Number of cylinders: 6
Top speed: 137mph (220kph)
Power: 120bhp @ 8000rpm
Transmission: 5-speed
Frame: Tubular double cradle

BELOW: The Kawasaki Z1300 of 1979 was the biggest capacity machine in Kawasaki's range and featured liquid-cooling, shaft drive and a six-cylinder engine across the frame. Six-cylinder engines never became really popular.

1981 KAWASAKI Z550 GP

The Z550 GP was the middleweight in Kawasaki's range of sports tourers, strategically placed between the bigger Z650F and the smaller Z400J of the same year's range. It bore more than a passing resemblance to both of these models with an angular gas

BELOW: Kawasaki redesigned its range of four-cylinder, four-strokes for the early 1980s and the Z550 of 1981 was the middleweight in the sports touring range.

tank that led into a similarly-shaped seat. Alloy wheels and twin front discs were all-typical of the time.

SPECIFICATION
Country of origin: JAPAN
Capacity: 553cc (33.73cu. in.)
Engine cycle: 4-stroke
Number of cylinders: 4
Top speed: 115mph (185kph)
Power: 54bhp @ 8500rpm
Transmission: 6-speed
Frame: Tubular double cradle

1984 KAWASAKI GPz550

The GPz models were refined and more sporting looking versions of the Z-series of a few years previously. Their design flowed more from the fairing into the gas tank and on through the seat and sidepanels. The

BELOW: The range of Z-prefixed models was redesigned to keep the machines current. The result was a range that looked like this GPz550 of 1984. The angular look was giving way to more flowing lines.

GPz550 was just one of a range that also included the GPz900R, GPz750R and the GPz305.

SPECIFICATION
Country of origin: JAPAN
Capacity: 553cc (33.73cu. in.)
Engine cycle: 4-stroke
Number of cylinders: 4
Top speed: 127mph (205kph)
Power: 65bhp
Transmission: 6-speed
Frame: Duplex tubular cradle

1984 KAWASAKI GPz900R

The GPz900R was Kawasaki's early 1980s contender in the competition between the various major Japanese manufacturers to offer the fastest, most powerful superbike. These bikes were all based around the successful and proven liquid-cooled, in-line four-cylinder engine in sports bike guise. The GPz900R engine was later overbored and stroked and used to power its successor the GPz1000R. The 'Big Four' Japanese manufacturers became embroiled in a race for performance which saw them updating their sports bikes annually in the quest to build the fastest and the best handling machines. This explains the rapid development of four-stroke, four-cylinder bikes.

SPECIFICATION
Country of origin: JAPAN
Capacity: 908cc (55.38cu. in.)
Engine cycle: 4-stroke
Number of cylinders: 4
Top speed: 156mph (252kph)
Power: 115bhp @ 9500rpm
Transmission: 6-speed
Frame: Tubular steel diamond

BELOW: In keeping with Kawasaki's practice, the sports tourers of all capacities bore a distinct resemblance to each other in both styling and paint schemes. The GPz900R of 1984 was the larger capacity model.

1990 KAWASAKI ZZR1100

The ZZR1100, or ZX11 in the USA, was the latest in a long line of powerful four-cylinder machines from Kawasaki, each one of which has been respected during its production run and many of which will become future classics in the way the first of the line, the Z1, is already doing. The ZZR1100 was the fastest production bike of its era and, although the medium capacity bikes are where the manufacturers have turned much of their attention for sports bikes, the biggest of their superbikes for the 1990s is, like its predecessors, well respected.

SPECIFICATION
Country of origin: JAPAN
Capacity: 1052cc (64.17cu. in.)
Engine cycle: 4-stroke
Number of cylinders: 4
Top speed: 175mph (282kph)
Power: 125bhp @ 10,000rpm
Transmission: 6-speed
Frame: Alloy box section

1991 KAWASAKI EN500

The EN500 is the antithesis of the ZZR1100 and one in a line of factory customs from Kawasaki. The pullback bars, two-level seat and slightly custom styling is a popular style for the USA and also sold elsewhere in the world. Other machines in this established style from the company have included the

ABOVE: The 1990 ZZR1100 was the largest capacity and fastest superbike of its era. It featured a 1052cc (64.17cu. in.) liquid-cooled, four-cylinder engine. The engine was almost entirely hidden behind the bodywork which had become popular at the time.

BELOW: The ZZR1100 stayed in production for several years although it was the subject of numerous engineering upgrades. It was also sold in different color schemes and with different graphics on its bodywork in each year of its manufacture. This is a 1996 model.

ABOVE: The Japanese manufacturers turned their attention away from the large capacity 'Superbikes' toward more medium-sized machines during the late 1980s and early 1990s. One such bike made by Kawasaki was the ZXR750 such as this 1991 model.

BELOW: The EN500 is Kawasaki's factory custom motorcycle, styled as it is with pullback handlebars, a stepped seat, curved tank and lines. These features give it the distinct appearance of an American machine, useful since the USA is a major export market.

Z440 LTD, the Z1000 LTD and the 450LTD. More recently there have been other cruisers including the VN1500 which is a large capacity heavyweight with styling that is similar to a Harley–Davidson.

SPECIFICATION
Country of origin: JAPAN
Capacity: 498cc (30.37cu. in.)
Engine cycle: 4-stroke
Number of cylinders: 2
Top speed: 99mph (160kph)
Power: 50bhp @ 1000rpm
Transmission: 6-speed
Frame: Tubular double cradle

1991 KAWASAKI ZXR750

The ZXR750 is a race-replica style of bike with many modern components including the monoshock rear suspension and upside-down front forks. A twin headlight fairing encloses the engine. The ZXR750R is a superbike racer which was manufactured in limited numbers.

SPECIFICATION
Country of origin: JAPAN
Capacity: 749cc (45.68cu. in.)
Engine cycle: 4-stroke
Number of cylinders: 4
Top speed: 155mph (250kph)
Power: 110bhp @ 10,500rpm
Transmission: 6-speed
Frame: Diamond section aluminum

1991 KAWASAKI ZEPHYR 550

The Zephyr was introduced in three capacities in a retro style reminiscent of the 1970s Z series of motorcycles. It was intended to appeal to riders returning to motorcycling or those who wanted a traditional looking machine rather than a race replica.

SPECIFICATION
Country of origin: JAPAN
Capacity: 553cc (33.73cu. in.)
Engine cycle: 4-stroke
Number of cylinders: 4
Top speed: 118mph (190kph)
Power: 50bhp @ 10,000rpm
Transmission: 6-speed
Frame: Steel tubular cradle

BELOW: The Kawasaki Zephyr was seen as a return to motorcycling's roots and the term 'Retro-bike' was soon coined. This 1991 model looked a lot like the Z900 of the late 1970s.

KRIEDLER

This German company started production in 1950 and concentrated on the manufacture of two-stroke 50cc (3.05cu. in.) mopeds. The machines are based around a pressed steel frame but are otherwise conventional. Kriedler machines have had some success in racing and broke world records for motor-

ABOVE: Completely typical of Kriedler's small capacity two-stroke machines is this 1967 Deluxe model. It is based around a 50cc (3.05cu. in.) engine and a pressed steel frame.

cycles of that category. At one time Kriedler were Germany's largest motorcycle manufacturer but the company went into decline to the extent that they went out of business in 1982. Their last production machines were sports and step-through mopeds, both were two-strokes.

1979 KRIEDLER RS FLORETT

The RS Florett is a sports moped with alloy wheels and a front disc brake. The two-stroke engine has a bore and stroke of 40 and 39.7mm enabling it to exceed 50mph (80kph). Its design is conventional although based around a pressed steel frame, much of which is hidden by the gas tank, seat and rear fender.

ABOVE RIGHT: The Kriedler RS Florett, seen on show in Brussels, Belgium, in 1977, was a sports moped based on the company's racing bikes which also displace 50cc (3.05cu. in.).

SPECIFICATION
Country of origin: GERMANY
Capacity: 49.9cc (3.04cu. in.)
Engine cycle: 2-stroke
Number of cylinders: 1
Top speed: 56mph (90kph)
Power: 6.25bhp @ 8500rpm
Transmission: 5-speed
Frame: Pressed steel

KTM

KTM is an acronym made up of the first letters of the surnames of its two founders and the town in Austria in which they are based: Kronreif, Trunkenpolz and Mattighofen respectively. The partnership was started in 1953 and the duo manufactured motorcycles with Rotax, Sachs and Puch engines. The company has made road machines but tends to specialize in off-road motorcycles.

1977 KTM 250 G56

This off-road bike from KTM is one of several models with a similar designation dependent on the capacity of its two-stroke engine. It is a competition-proven product. The company's riders won the 250 World Moto-X Championship in 1974. The other machines of this type produced between the mid-1970s and 1980s include the 400 GS and the 350 GS.

SPECIFICATION
Country of origin: AUSTRIA
Capacity: 246cc (15cu. in.)
Engine cycle: 2-stroke
Number of cylinders: 1
Top speed: n/a
Power: 34bhp @ 7400rpm
Transmission: 6-speed
Frame: Tubular cradle

ABOVE: KTM are noted for the production of off-road competition machines such as this pair of enduro bikes of 600 and 300cc respectively (36.60 and 18.30cu. in.).

BELOW: The LC4 built by KTM in 1991 was a state-of-the-art off-road enduro bike. It featured long travel suspension and a four-stroke, single-cylinder engine in a cradle frame.

1991 KTM 600 LC4

The 600 LC4 was the largest of the enduro bikes manufactured by KTM in the early 1990s. It features a four-stroke engine, although the smaller capacity machines which otherwise bear a close resemblance to it have two-stroke engines, namely the 300 Enduro and the 125 Enduro. Long suspension travel, high-tech forks, swing arms and disc brakes are typical of 1990s off-road competition machines.

SPECIFICATION
Country of origin: AUSTRIA
Capacity: 553cc (33.73cu. in.)
Engine cycle: 4-stroke
Number of cylinders: 1
Top speed: n/a
Power: 52bhp @ 8500rpm
Transmission: 5-speed
Frame: Chrome molybdenum tubular

L

LAMBRETTA

The Italian company Innocenti of Milan started manufacturing scooters in the years immediately following World War II. Italy lay in ruins due to the ravages of the war. The Germans had fought the Allies for much of Italian soil. Allied bombing had destroyed many factories and towns, and the Germans had destroyed road and rail links as they retreated slowly north in the face of the Allied advance. Innocenti had been an established company with a long history in manufacturing. With American money for reconstruction they sought to

BELOW: The Scooter, primarily a product of postwar Italy, was revolutionary in both its method of construction and its appearance. It was its futuristic appearance which made it popular and fashionable around Europe. This is a 1960 Lambretta – clean and stylish.

meet the immediate need for mass transportation and intended it as a temporary way back into industrial production. The result was the Lambretta which had been designed by a team headed by Pier Luigi Torre. The motorcycle took its name from the River Lambret that flowed past the factory. The early models relied on Innocenti's experience with steel tubing for industrial applications. It was only later in their history that Lambrettas had enclosed engines. The company intended these machines as practical utility transport and as such they were completely successful. In fact, demand outstripped supply. They also became style items offering a futuristic look and symbolizing a new beginning for the world in the postwar era.

1952 LAMBRETTA MODEL D

Lambretta felt pressured by the public into fitting bodywork to their rear-engined machines to compete with rival Vespa. To ensure they were heading the right way in the eyes of their customers they offered the

Model D, which was an unenclosed scooter, and the LD, which was a fully enclosed model. The enclosed model sold in such greater numbers that Innocenti stopped making unenclosed machines after 1958.

SPECIFICATION
Country of origin: ITALY
Capacity: 123cc (7.50cu. in.)
Engine cycle: 2-stroke
Number of cylinders: 1
Top speed: 45mph (72kph)
Power: 5bhp @ 4600rpm
Transmission: 4-speed
Frame: Steel tube spine

1954 LAMBRETTA LD

The LD was introduced in 1952 as a fully enclosed version of the Model D, a 125cc (7.62cu. in.) scooter based around the steel tubular spine of the Innocenti design. It and Vespa's products were the result of the postwar reconstruction of Italy and what is accepted as Italy's industrial revolution. The Model LD outsold the exposed Model D by

ABOVE: Lambretta initially built scooters with and without enclosed engines. The LD had bodywork but the L did not. It was the LD, such as this 1954 model, that caught on.

a considerable margin so that in 1958 the D was dropped in favor of the LD.

SPECIFICATION
Country of origin: ITALY
Capacity: 125cc (7.62cu. in.)
Engine cycle: 2-stroke
Number of cylinders: 1
Top speed: 45mph (72kph)
Power: 5bhp @ 4600rpm
Transmission: 3-speed
Frame: Steel tubular spine

1980 LAMBRETTA GRAND PRIX 200

Lambrettas were built under license by Serveta in Spain and by Scooters India Ltd. in Lucknow, India. Innocenti stopped the production of Lambrettas in the 1960s, when they were taken over by British Leyland, the car manufacturer, to produce an Italian version of the Mini car. The Grand Prix 200 was still being made in India long after that.

SPECIFICATION
Country of origin: INDIA
Capacity: 198cc (12.07cu. in.)
Engine cycle: 2-stroke
Number of cylinders: 1
Top speed: 70mph (112kph)
Power: 12.4bhp @ 6300rpm
Transmission: 4-speed
Frame: Steel tube spine

BELOW: This 1969 Lambretta Grand Prix was a stylized scooter. Seating for two, some weather protection and ease of riding were some of the reasons it was popular.

LAVERDA

The family-owned industrial group that produced Laverda motorcycles was started in 1873 when Pietro Laverda started an agricultural machinery factory at Breganze in Italy. Moto Laverda was founded in 1949 and started motorcycle production with a 74cc (4.51cu. in.) overhead-valve motorcycle engine fitted into a pressed steel frame. This was one of a series of small displacement motorcycles made by the Italian company. In 1968 the company produced a 650cc (39.65cu. in.) machine along Honda lines (it was closely based on Honda's CB72 250cc twin) and when its displacement was increased to 750cc (45.75cu. in.) exports began. In 1973 a new Laverda factory was opened and new models soon rolled off its production lines. Most notable was the double overhead camshaft triple of 981cc (59.84cu. in.) displacement which was followed by a 1200cc (73.20cu. in.) variant. The company also produced smaller capacity motorcycles in both roadgoing and off-road trim. The company stopped motor-

cycle production in 1988 although it was later restarted by another company known as Nuova Moto Laverda.

1973 LAVERDA 750SF

The 750SF engine was originally introduced in the late 1960s as a 650cc (39.65cu. in.) parallel-twin engine but was redesigned in

ABOVE: The Laverda 668 of 1996 is an air- and oil-cooled four-stroke, twin-cylinder sports bike of 668cc (40.74cu. in.) displacement as indicated by the model's designation.

BELOW: The Laverda 750SF was typical of Italian sports bikes of the time. Seen here is the 1973 model with the four leading shoe drum brake, replaced by discs in 1975.

1972 when it acquired the sharply angled cooling fins running to the top of the cam-box. The motorcycle, while not the fastest production machine of its day, had a certain aggressive and yet attractive style through the use of a long sleek tank, flat handlebars and the sporting riding position. New to the machine for 1975 were the twin front brake discs which were in place of a four leading shoe front drum and a new headlamp. The previous year had also seen the 750SF fitted with Dellorto carburetors.

SPECIFICATION
Country of origin: ITALY
Capacity: 744cc (45.38cu. in.)
Engine cycle: 4-stroke
Number of cylinders: 2
Top speed: 117.5mph (189kph)
Power: 65bhp @ 7000rpm
Transmission: 5-speed
Frame: Tubular spine

1974 LAVERDA 3CL

The Laverda 3CL was a sports bike powered by a four-stroke, double overhead camshaft engine from which the Jota would ultimately evolve as a faster race-inspired version. The company at this time was producing a total of around 8000 motorcycles per year, of which the 3CL formed a significant portion.

ABOVE: The Laverda 3CL of 1974, such as this one, was the forerunner of the Laverda Jota, a motorcycle which was destined to become a legend. The 3CL featured a double overhead camshaft, triple cylinder engine of 981cc (59.84cu. in.). For use in the Laverda Jota, a three-cylinder engine of the same displacement was utilized but it was given a higher state of tune.

SPECIFICATION
Country of origin: ITALY
Capacity: 981cc (59.84cu. in.)
Engine cycle: 4-stroke
Number of cylinders: 3
Top speed: 130mph (209 kph)
Power: 81bhp @ 7200rpm
Transmission: 5-speed
Frame: Duplex cradle

1980 LAVERDA JOTA 1000

The Laverda Jota was a motorcycle which utilized a three-cylinder, double overhead camshaft engine and it quickly became renowned as one of the superbikes of the 1970s. A major selling point was that the Jota was the fastest motorcycle available on its introduction. Its powerful engine was paired with a chassis which possessed good handling characteristics. The Jota was state-of-the-art at the time of its introduction with Brembo disc brakes front and rear, Ceriani forks, fully adjustable handlebars, cast alloy wheels and sporting lines.

SPECIFICATION
Country of origin: ITALY
Capacity: 981cc (59.84cu. in.)
Engine cycle: 4-stroke
Number of cylinders: 3
Top speed: 140mph (225kph)
Power: 85bhp @ 7600rpm
Transmission: 5-speed
Frame: Duplex cradle

LEFT: The Laverda Jota was a fast, race-track inspired triple and, as this 1980 model illustrates, featured Brembo discs, Ceriani forks and cast alloy wheels. These were just two of the features which made it a state-of-the-art motorcycle at the time of its introduction.

ABOVE: In 1980 the 1200cc triple was Laverda's largest displacement engine and was used in three models of motorcycle that the company produced, including this, the 1200TS, a sports tourer with a fairing.

1980 LAVERDA 1200

The 1200 was the largest displacement bike in Laverda's range at the beginning of the 1980s and was available in three forms: the 1200, the Mirage and the 1200TS. The latter model was a luxury version that was fitted with a fairing and partial engine enclosure while the Mirage was a faster version capable of 137mph (220kph).

SPECIFICATION
Country of origin: ITALY
Capacity: 1116cc (68.07cu. in.)
Engine cycle: 4-stroke
Number of cylinders: 3
Top speed: 130mph (209kph)
Power: 87bhp @ 7800rpm
Transmission: 5-speed
Frame: Duplex tubular cradle

1991 LAVERDA NAVARRO

The Navarro was manufactured by the reformed Laverda company. It was a small capacity sports bike based around a liquid-cooled, two-stroke engine. Despite its small displacement, the Laverda Navarro featured race bike styling and components such as exhausts, disc brakes and alloy wheels.

SPECIFICATION
Country of origin: ITALY
Capacity: 124.63cc (7.60cu. in.)
Engine cycle: 2-stroke
Number of cylinders: 1
Top speed: 99mph (160kph)
Power: 22bhp @ 10,500rpm
Transmission: 7-speed
Frame: Steel trellis

LEVIS

This British firm dated from 1911 and in its early years was best known for the production of two-stroke motorcycles. The company had some racing success with such machines during the early 1920s. The company's roadgoing models reflected their enthusiasm for two-strokes and a six port 247cc (15.06cu. in.) machine as well as the 247cc (15.06cu. in.) Model Z were included in the Levis range into the 1930s. These two models were listed in the catalog alongside a variety of medium capacity four-strokes. The two-strokes were dropped in the early 1930s although one was later reintroduced, as was a 498cc (30.37cu. in.) four-stroke with total loss lubrication. The range saw alterations through the remainder of the 1930s and production stopped in 1940, signaling the end of the marque.

1939 MODEL D SPECIAL LEVIS

The Model D appeared as a new model in the Levis range for 1934 and ran on for several years in essentially the same form. It featured an overhead-valve, single-cylinder engine fitted into a completely orthodox frame and cycle parts. In the later years of its production, plunger rear suspension was available with a hydraulic pipe running between the two sides of the frame to balance the load and provide a degree of damping.

SPECIFICATION
Country of origin: GREAT BRITAIN
Capacity: 498cc (30.37cu. in.)
Engine cycle: 4-stroke
Number of cylinders: 1
Top speed: n/a
Power: n/a
Transmission: n/a
Frame: Tubular steel

BELOW: A 1939 Levis 500 D Special. This was an orthodox motorcycle powered by a single-cylinder engine. During its production run it was upgraded with plunger suspension.

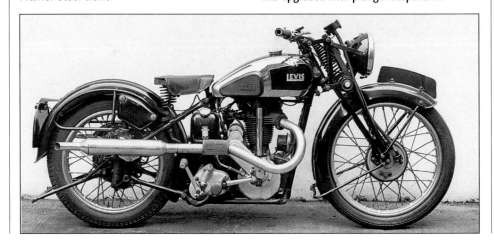

M

MABECO

This German company was in business between 1923 and 1928 and during that period produced copies of the American Indian Scout. These were built by Max Bernhardt in Berlin with engines made by Siemens and Halske. The bikes were even red in color like Indians so the Springfield, Massachusetts, company went to court and Mabeco went into liquidation. Later, however, they restarted motorcycle production as Mabeco–Werke GmbH, yet still with Siemens and Halske as major shareholders. Later they built an Italian Garelli under license from the Italian company.

MAICO

Otto and Wilhelm Maisch founded this company in 1926. Their first motorcycle was a 98cc (6cu. in.) roadgoing machine with an Ilo engine. During World War II the company moved to Pfäffingen where it manufactured parts for the Luftwaffe. After the war the company was again producing motorcycles by 1950. The postwar range included a 400cc (24.40cu. in.) twin and

BELOW: The Maico company latterly specialized in the production of off-road motorcycles such as this 1977 machine known as the Maico Sporting. It was machines such as this that were popular and successful in the USA.

touring scooters. Later the company moved towards specializing in off-road motorcycles and its products were widely exported, especially to the USA. Maico motorcycles have been successful in both 250 and 500 Motocross World Championships. Production of Maicos stopped in 1983.

1977 MAICO MD 125 SUPERSPORT

The MD 125 Supersport was a traditionally-designed, small capacity motorcycle aimed at commuters and those wanting basic motorcycling. By dint of its displacement and size the machine was cheap to buy and cheap to run.

SPECIFICATION
Country of origin: GERMANY
Capacity: 124cc (7.56cu. in.)
Engine cycle: 2-stroke
Number of cylinders: 1
Top speed: 76mph (123kph)
Power: 16bhp @ 7800rpm
Transmission: 6-speed
Frame: Tubular steel

MALAGUTI

Malaguti is one of Italy's largest moped manufacturers and was founded in 1937 by the Malaguti family. The company is based in Bologna, northern Italy, and started out by making mopeds fitted with 38cc (2.31cu. in.) Garelli engines. Other makes of small capacity engines including Franco Morini and Sachs have been used. The company now concentrates on mopeds.

1980 MALAGUTI CAVALCONE

This was a small capacity off-road style of bike fitted with a Morini engine and was one of 12 models in the 1980 range. One of this range was a moped designed for children and another included a model with pushbike pedals to allow its use by younger teenagers in certain countries.

SPECIFICATION
Country of origin: ITALY
Capacity: 49.9cc (3.04cu. in.)
Engine cycle: 2-stroke
Number of cylinders: 1
Top speed: 50mph (80kph)
Power: 6.5bhp @ 8500rpm
Transmission: 4-speed
Frame: Duplex cradle

BELOW: Malaguti use proprietary engines in their range of small-capacity, off-road machines. The Malaguti 76 Super Cross (shown) is powered by a Morini single-cylinder engine, for example, and was one of a range of lightweight off-road motorcycles for this era.

1991 MALAGUTI RST 50

By 1991 Malaguti's range included a number of models which were all based around the same basic two-stroke engine. The RST 50 was a small sports motorcycle while the MRX 50 was an enduro-type of off-roader, and the Top 50 was a step-through style moped. Although all of the engines were of the same basic type, the machines utilized different types of frame in order to suit their different styles: a box section cradle for the sports model; a tubular cradle for the MRX 50; and a pressed steel step-through for the Top 50. The cycle parts were all styled in order to suit the particular type of moped in question.

SPECIFICATION
Country of origin: ITALY
Capacity: 49.9cc (3.04cu. in.)
Engine cycle: 2-stroke
Number of cylinders: 1
Top speed: 24.8mph (40kph)
Power: 1.5bhp @ 6200rpm
Transmission: 3-speed
Frame: Box section cradle

MARS

This German company, which was based in Nuremberg, was founded in 1903 and produced motorcycles with German Fafnir and Swiss Zedel engines until the outbreak of World War I. After this a Herr Franzenburg designed a machine with a welded and riveted box section frame. It was fitted with a side-valve flat-twin Maybach engine of 956cc (58.31cu. in.) displacement. The bike had other unorthodox features including a hand-starter. It sold well until the company fell on hard times during the period of rampant inflation that troubled the German economy during the 1920s. Production was restarted by two employees of the company but they were unable to use the Mars tradename. From then on machines were known as MA and featured a number of proprietary engines. In the years leading up to World War II the firm produced machines using Sachs 147, 174 and 198cc (8.96, 10.61 and 12.07cu. in.) engines. Production ended in 1957 when the company closed down.

RIGHT: The Mars 175 was a two-stroke motorcycle manufactured by the company in the years after World War II. This one was made in 1954 soon before production stopped.

MATCHLESS

This English company was founded in 1899 and for its first years used a large variety of proprietary engines from the likes of JAP and De Dion. Harry and Charlie Collier had considerable Isle of Man TT successes on Matchless machines in the earliest years of the event. Matchless bought AJS in 1931 and by 1935 had introduced models that were of the same style but sold under both marques' badges. Later they became more entwined when bikes made by each com-

ABOVE: Matchless were amongst the pioneers of the British motorcycle industry. Like others at the time they used proprietary engines fitted into frames that show the machine's bicycle origins, as in this 1902 model.

pany were the same except for the badges. At the beginning of the 1930s the Matchless company had a range of singles of varying capacities, and a V-twin intended for sidecar duties. The company conducted experiments with a couple of V configuration engines – a V-twin and a V-4 that was

essentially a pair of V-twins which had been joined together.

The G3 Clubman was introduced in 1935 and set the style for the Matchless range from that point onward. In 1937 the Matchless acquired Sunbeam and became the AMC (the Associated Motorcycles Company). World War II was looming by 1938 and during the war years the company built around 80,000 Matchless 350cc (21.35cu. in.) G3 and G3L models. The G3 had girder forks. It was superseded in 1941 by the G3L which had telescopic forks.

In the postwar years Matchless and AJS models were identical although the company persevered with both brands, even advertising them separately. AMC later took over Francis–Barnett, James, Norton and Indian of Springfield, Massachusetts, USA. The company went into decline soon afterward and later became part of a group owned by Dennis Poore. He transferred production to Andover in Hampshire, England, and continued production of Norton and AJS machines. Matchless, as a marque, gradually became history.

1930 MATCHLESS SILVER ARROW

This machine featured an unorthodox engine and was an attempt by Matchless to

ABOVE: The Matchless Silver Arrow of 1930 was a sophisticated touring machine that used an unusual side-valve V-twin engine in an otherwise conventional motorcycle. It was not a success in terms of sales.

sell a sophisticated touring machine. It was powered by a side-valve 26° V-twin that was cast as a single block and had only one exhaust pipe which emerged from the front. As a result the overall appearance was of an overlong single. The camshaft was gear driven and enclosed in the right crankcase. The motorcycle had chain primary and final drives and the whole was fitted into a frame with pivoted-fork rear suspension and girder forks. The Silver Arrow did not sell particularly well and was later replaced by the Silver Hawk with a V4 engine which had a displacement of 593cc (36.17cu. in.).

SPECIFICATION
Country of origin: GREAT BRITAIN
Capacity: 394cc (24.03cu. in.)
Engine cycle: 4-stroke
Number of cylinders: 2
Top speed: n/a
Power: n/a
Transmission: 3-speed
Frame: Tubular steel

1931 MATCHLESS MODEL B SILVER HAWK

This machine was unveiled at the beginning of the 1930s and listed in the Matchless range as the 34/B. It featured an overhead camshaft engine of 592cc (36.11cu. in.) in a V-four configuration. The engine was essentially two 26° Silver Arrow V-twin engines combined into a greater capacity V-four. It featured a three-bearing, two-throw crankshaft housed within a one-piece block, capped with a one-piece cylinder head. A single camshaft was driven by shaft and bevels and a skew gear on the camshaft drove a shaft which in turn drove the dynamo and distributor for the coil ignition system. However, the Silver Hawk did not achieve much success as it was an expensive machine launched during the first years of the Great Depression in the early 1930s.

SPECIFICATION
Country of origin: GREAT BRITAIN
Capacity: 592cc (36.11cu. in.)
Engine cycle: 4-stroke
Number of cylinders: 4
Top speed: 80mph (128kph)
Power: 26bhp
Transmission: 4-speed
Frame: Spring frame

ABOVE: The postwar Matchless single was a
developed version of the G3/G3L from the
years of World War II. It featured the
telescopic forks of the G3L and acquired
swingarm rear suspension.

1940 MATCHLESS G3

AMC produced, as Matchless and AJS models, in excess of 80,000 G3 and G3L motorcycles for the British and Allied armies during World War II. The difference between the two models was that the G3L had a lighter frame and telescopic forks. The G3L was introduced in 1941 and its superior handling compared to the earlier model soon endeared it to many Despatch riders.

SPECIFICATION
Country of origin: GREAT BRITAIN
Capacity: 347cc (21.16cu. in.)
Engine cycle: 4-stroke
Number of cylinders: 1
Top speed: 70mph (112kph)
Power: 16bhp
Transmission: 4-speed
Frame: Tubular steel

ABOVE: The postwar 347cc (21.16cu. in.) known as the G3L Matchless single benefited from the telescopic forks introduced during World War II on the Despatch rider's G3.

1951 MATCHLESS G80S

The G80 was a large capacity single with an overhead-valve configuration engine. It was one of a range of singles produced by the company in the years following World War II.

SPECIFICATION
Country of origin: GREAT BRITAIN
Capacity: 497cc (30.31cu. in.)
Engine cycle: 4-stroke
Number of cylinders: 1
Top speed: n/a
Power: n/a
Transmission: 4-speed
Frame: Tubular cradle

ABOVE: The G80S was a 497cc (30.31cu. in.) single-cylinder motorcycle with a swingarm frame, telescopic forks and drum brakes.

1955 MATCHLESS G9 CLUBMANS

In the same way that the company produced a range of singles, it also produced a range of twins. Both the singles and the twins were used extensively for competitive motorcycling: the singles competed in off-road events while the twins were used for road and circuit racing.

SPECIFICATION
Country of origin: GREAT BRITAIN
Capacity: 498cc (30.37cu. in.)
Engine cycle: 4-stroke
Number of cylinders: 2
Top speed: n/a
Power: n/a
Transmission: 4-speed
Frame: Tubular steel

ABOVE: Matchless built twin-cylinder machines alongside their singles. This 1955 G9 was aimed at club racers.

1957 MATCHLESS G45

This was a racing bike developed by the AMC factory for their riders to compete in various TT races. The first G45s were developed in time for the 1952 season and were based on a parallel-twin engine. The engine featured a forged steel one-piece crankshaft, alloy barrels, twin Amal carburetors, a Lucas racing magneto and a four-speed Burman racing gearbox.

SPECIFICATION
Country of origin: GREAT BRITAIN
Capacity: 498cc (30.37cu. in.)
Engine cycle: 4-stroke
Number of cylinders: 2
Top speed: 120mph (193kph)
Power: 48bhp @ 7200rpm
Transmission: 4-speed
Frame: Tubular cradle

1958 MATCHLESS G11 CSR

The G11 CSR model was developed for the American market which was a strong export destination for British-manufactured motorcycles during the 1950s and 1960s. As a result it was styled in a fashion appropriate to that market. It drew some of its inspiration from dirt bikes of the era and was an early example of what came to be known as the 'street scrambler'.

SPECIFICATION
Country of origin: GREAT BRITAIN
Capacity: 593cc (36.17cu. in.)
Engine cycle: 4-stroke
Number of cylinders: 2
Top speed: n/a
Power: 40bhp @ 6000rpm
Transmission: 4-speed
Frame: Duplex cradle

1961 MATCHLESS G50 CSR

This motorcycle was built in very limited numbers to enable the model G50 to become eligible for motorcycle competition in the USA. It was simply a combination of the rolling chassis of the CSR twin with an

BELOW: The G50 CSR of 1961 was built for racing and in limited numbers just to ensure that it was homologated for competition in American motorcycle sport.

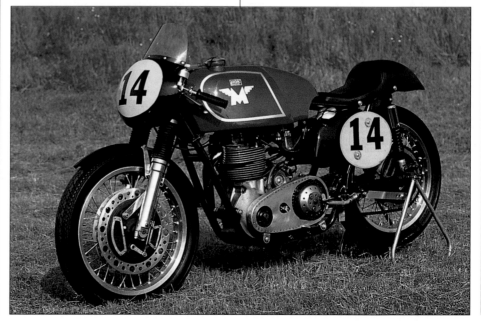

overhead camshaft, G50 single-cylinder engine fitted.

SPECIFICATION
Country of origin: GREAT BRITAIN
Capacity: 498cc (30.37cu. in.)
Engine cycle: 4-stroke
Number of cylinders: 1
Top speed: 115mph (185kph)
Power: 46bhp @ 7200rpm
Transmission: 4-speed
Frame: Tubular cradle

MEGOLA

This short-lived German maker produced what are generally considered to have been some of the world's most unorthodox motorcycles ever. The company started production in 1921 and stopped in 1925 but in that time produced motorcycles with five-cylinder radial engines built into the front wheel. They were clearly based on radial aeroplane engines but did not feature components such as a clutch or a gearbox. Fritz Cockerell designed the air-cooled, side-valve radial engine which was started by the rider kicking the front wheel round while the machine was on its main stand. The name Megola was an acronym derived

ABOVE: Megola built unorthodox motorcycles but had racing success due to good handling characteristics. This 1922 model won the German Championship in 1924.

from the names of the three founders; Meixner, Cockerell and Landgraf, although certain latitude was taken with Cockerell's name. The company made around 2000 motorcycles in total.

1922 MEGOLA RADIAL

The unusual engine application meant that the changing of gear ratios could only be accomplished through the use of different diameter wheels. This was done for racing purposes because, although of unorthodox appearance, the Megola radial had a low centre of gravity and so handled well. Front suspension was by means of semi-elliptic springs in a substantial cradle while some models had semi-elliptic rear springs too. There were other clever features of the design. It was possible to dismantle the engine cylinders without removing the spokes of the wheel. Also, the box section frame contained the main fuel tank although fuel had to be decanted into a smaller fork-mounted tank at intervals.

Sports Megolas featured saddles while the roadgoing models had bucket seats.

SPECIFICATION
Country of origin: GERMANY
Capacity: 637.2cc (38.86cu. in.)
Engine cycle: 4-stroke
Number of cylinders: 5
Top speed: 52mph (85kph)
Power: 14bhp @ 3600rpm
Transmission: Single-speed
Frame: Box section

BELOW: The Megola featured a five-cylinder, radial engine which was unusually positioned in the front wheel of the motorcycle.

MINARELLI

This company was originally called FBM and was a collaboration between Franco Morini and Vittorio Minarelli. Between 1951 and 1957 they produced a 125cc (7.62cu. in.) two-stroke single, which was called the Gabbiano. In 1956 FBM started supplying its own engines to other companies. Franco Morini left in 1957 and FBM became known as FB Minarelli. The company today concentrates in the main on supplying engines to other companies.

BELOW: Minarelli supplied its own engines to other motorcycle manufacturers including the British trials bike company DOT, as seen on this early 1960s example.

MONARCK

This Swedish company was one of four using the tradename 'Monarch' at various times (the spelling varies). The others were in England, Japan and America, although all three existed for less than a decade. The Swedish firm on the other hand became more established from its beginnings in 1920 under the Esse brandname. Its first machines were all fitted with Blackburne engines and by the mid-1930s the company was manufacturing Ilo powered lightweights. A Husqvarna/Albin engine was used for an army bike during the years of World War II despite Sweden's neutrality. In the postwar years a range of two-stroke bikes, in the main using Ilo engines, were

produced, and by the 1950s the concern was Sweden's biggest motorcycle manufacturer. In 1960 Monarck absorbed NV which was another Swedish manufacturer, named after Nymans Verkstäder, and became known as MCB. From that point onward the company made lightweight machines powered by Morini and Sachs engines until the cessation of production in 1975.

MONET GOYON

This French company, founded in 1917, were based in Macon and used proprietary engines built by Swiss MAG and English Villiers. In the early 1930s they produced a 344cc (20.98cu. in.) side-valve unit single which was fitted into a traditional motorcycle with a rigid frame and girder forks. After World War II a range of two-strokes were produced that varied in capacity from 98cc (5.97cu. in.) to 232cc (14.15cu. in.). In the latter years – the company stopped production in 1957 – Monet Goyon manufactured a scooter known as the Starlet.

ABOVE: Monet Goyon were a French manufacturer who used proprietary engines in their machines, including this 1930s model.

MONTESA

Montesa was the first motorcycle manufacturer founded in Spain. Its origins are in a workshop founded by Pedro Permanyer-Puigjaner in 1941. A new modern plant was opened in 1962 where mass production of engines, outboard motors and motorcycles took place. Montesa specialize in off-road motorcycles for trials and motocross and their products have an enviable international reputation.

1977 MONTESA COTA 348

The Montesa Cota is a competent trials bike for off-road use and features a short wheelbase, high ground clearance and an engine that delivers its torque at low revs. The Cota

ABOVE: A 1976 Montesa Cota 348. This was one of the Spanish company's specialized off-road trials competition machines.

348 is one of a range of models so named. Others include the 247, 74T and 25.

SPECIFICATION
Country of origin: SPAIN
Capacity: 305.8cc (18.65cu. in.)
Engine cycle: 2-stroke
Number of cylinders: 1
Top speed: n/a
Power: n/a
Transmission: 6-speed
Frame: Tubular cradle

1991 MONTESA COTA 335

The successful Cota series of trials motorcycles was retained in production, although upgraded as trial bike technology progressed. The 1991 model had progressed from a swinging arm with twin shock absorbers and rear suspension assembly to a monoshock. Similarly drum brakes had been succeeded by discs, front and rear. However, other features remained the same – a 21in. (530mm) diameter front wheel and an 18in. (450mm) rear, for example.

SPECIFICATION
Country of origin: SPAIN
Capacity: 327.8cc (19.99cu. in.)
Engine cycle: 2-stroke
Number of cylinders: 1
Top speed: n/a
Power: 17.5bhp
Transmission: 6-speed
Frame: Duplex tubular cradle

MORBIDELLI

This Italian motorcycle was produced only in very limited numbers and was the project of Giancarlo Morbidelli, a wealthy Italian industrialist and motorcycle enthusiast. The motorcycle was truly one of the world's most exotic machines; included in the initial expensive purchase price was a service contract that included airfreighting the motorcycle back to its Italian factory for servicing.

1994 MORBIDELLI V8

The V8 featured contemporary sports bike styling, with an integral seat, fairing and tank unit. It was designed by the noted Italian car stylist Pininfarina and featured a tubular space frame from which the V8 engine was suspended. The engine was of a water-cooled, double overhead camshaft configuration. The remainder of the motorcycle used quality proprietary components such as Marvic spoked alloy wheels and Brembo brake discs and calipers.

SPECIFICATION
Country of origin: ITALY
Capacity: 847cc (51.66cu. in.)
Engine cycle: 4-stroke
Number of cylinders: 8
Top speed: 150mph (241kph)
Power: n/a
Transmission: n/a
Frame: Tubular space frame

BELOW: The Morbidelli V8 of 1994 is one of the world's most exotic motorcycles and one designed for the wealthy. It is powered by a V8 engine with a displacement of 847cc (51.66cu. in.). The complex engine was liquid-cooled with double overhead camshafts. The machine was made in tiny quantities.

LEFT: While Motobecane did produce a larger capacity machine known as a Motoconfort they also produced this mini-bike so named. It uses a similar engine and transmission to the step-through model but has smaller wheels.

1977 MOTOCONFORT 350

In 1970 Motobecane commenced production of a 125cc two-stroke twin. This was later followed by a 350cc triple and then, in 1977, a 500cc triple, a motorcycle which incorporated electronic fuel injection. The 350 model was named after the other company with which Motobecane had combined in 1930.

SPECIFICATION
Country of origin: FRANCE
Capacity: 349cc (21.28cu. in.)
Engine cycle: 2-stroke
Number of cylinders: 3
Top speed: 102.5mph (165kph)
Power: 38bhp @ 7200rpm
Transmission: 5-speed
Frame: Steel tubular cradle

1991 MOTOBECANE ZX50 TRAIL

By 1991 Motobecane were selling their products labeled as MBK – a phonetic interpretation of Motobecane. The ZX50 Trail was a small capacity trail bike with styling entirely typical of dirt bikes of the era.

SPECIFICATION
Country of origin: FRANCE
Capacity: 49.9cc (3.04cu. in.)
Engine cycle: 2-stroke
Number of cylinders: 1
Top speed: n/a
Power: 4.8bhp @ 7000rpm
Transmission: Automatic
Frame: Tubular cradle

MOTOBECANE

Two motorcycle companies, Motobecane and Motoconfort, started in business in France during the early 1920s. In 1930 they combined and thrived and by the 1980s the company was the world's largest manufacturer of mopeds and bicycles. In the years between the two World Wars they produced 500 and 750cc (30.50 and 45.75cu. in.) fours but since the end of World War II they have concentrated on small capacity two-strokes. In 1949 Motobecane launched the Mobylette, of which in excess of 11 million have been manufactured. The company has a factory at Saint Quentin, France, and an engine plant in Pantin, France. The company produces its own two-stroke engine and also offers step-through designs of moped as well as a range of small capacity motorcycles.

License-building has been permitted in the following countries: Spain, Iran, Zaire and Morocco.

SPECIFICATION
Country of origin: FRANCE
Capacity: 49.9cc (3.04cu. in.)
Engine cycle: 2-stroke
Number of cylinders: 1
Top speed: 25mph (40kph)
Power: 1.8bhp @ 4500rpm
Transmission: Single-speed
Frame: Steel step-through

1970 MOBYLETTE 40 VS

The Mobylette is entirely conventional as a step-through style of moped designed purely as a cheap, economical and basic form of transport. These attributes are the keys to its success and it has been produced in enormous numbers since its introduction.

RIGHT: The Mobylette Commuter Deluxe of 1974 is simply the most basic form of step-through moped manufactured. It features a tiny engine with a single-speed transmission.

MOTO GUZZI

In 1921 Carlo Guzzi and Giorgio Parodi founded the Moto Guzzi company to produce motor-cycles at Mandello Del Lavio, Italy. Their first machines featured a 499cc (30.43cu. in.) single-cylinder, inlet-over-exhaust valve engine with a horizontal forward-facing cylinder. The same basic design was retained when the company produced overhead-valve and overhead camshaft designs of a similar capacity and later they made a 250cc (15.25cu. in.) version. The company experimented with multi-cylindered racing machines during the 1930s in addition to supercharged motorcycles. The company manufactured motorcycles for the Italian Army during World War II and then returned to manufacturing racing and road bikes when the war ended. One of the former was the eight-cylinder, 498cc (30.37cu. in.) double camshaft, water-cooled works motorcycle of 1955.

In the years which followed World War II Moto Guzzi commenced production of a comprehensive range of smaller capacity machines, both scooters and motorcycles. They opted out of racing in the late 1950s on the grounds of cost. The company was administered by the state in the late 1960s when it was on the verge of bankruptcy.

BELOW: A 1952 Moto Guzzi 250 Airone. This machine and the Falcone were 1950s classics, manufactured by the Italian firm who were among the first to use a wind tunnel in design.

During the 1970s the company became part of the Alessandro de Tomaso group which also controlled Benelli and co-operated closely with Motobecane of France. Lino Tonti designed the transverse V-twin engined machines which were included in Moto Guzzi's range from 1969 onward.

1979 MOTO GUZZI 850 LE MANS II

The Le Mans II was the successor to the Le Mans of 1976 and named after the marque's sporting success at the Bol d'Or endurance race held at Le Mans, France. The engine was not originally developed for a motorcycle but for a military cross-country vehicle. However, its smoothness, as a

ABOVE: The Moto Guzzi Le Mans II was so named after a win at the Bol d'Or Endurance Race at Le Mans in France. It used Lino Tonti's transverse V-twin engine.

result of being a 90° V-twin, made it suitable for two-wheeled applications.

SPECIFICATION
Country of origin: ITALY
Capacity: 844cc (51.48cu. in.)
Engine cycle: 4-stroke
Number of cylinders: 2
Top speed: 125mph (201kph)
Power: 80bhp @ 7300rpm
Transmission: 5-speed
Frame: Duplex tubular cradle

1981 MOTO GUZZI V50 MONZA

The V50 was the Moto Guzzi machine designed to fill the gap between its small capacity two-stroke singles and its big four-stroke V-twins. Its engine was a redesigned version of the transverse V-twin used in the larger models and the motorcycle was built at the Innocenti plant. Its construction featured electric start, shaft drive, alloy wheels and Brembo brakes.

SPECIFICATION
Country of origin: ITALY
Capacity: 490.2cc (29.90cu. in.)
Engine cycle: 4-stroke
Number of cylinders: 2
Top speed: 108mph (175kph)
Power: 48bhp @ 7600rpm
Transmission: 5-speed
Frame: Tubular cradle

1991 MOTO GUZZI QUOTA 1000

This motorcycle was Moto Guzzi's machine of the early 1990s, inspired by desert races. It featured the proven transverse V-twin engine, favored by Moto Guzzi, protected by a bash plate but set in a frame with enhanced ground clearance, a Marzocchi

ABOVE: Desert races such as the Paris-Dakar inspired many makers to build machines that resembled those race bikes. The Quota 1000 was Moto Guzzi's response and retained the transverse V-twin that is their hallmark.

monoshock rear suspension assembly, and telescopic forks. Disc brakes front and rear and shaft drive were standard features, as was a mini fairing with twin headlamps.

SPECIFICATION
Country of origin: ITALY
Capacity: 948.8cc (57.87cu. in.)
Engine cycle: 4-stroke
Number of cylinders: 2
Top speed: 119mph (192kph)
Power: 75bhp @ 7400rpm
Transmission: 5-speed
Frame: Box section cradle

1991 MOTO GUZZI 750T TARGA

The 750T is a sports bike and shares its styling with the Le Mans but features a smaller capacity engine denoted by its model designation. The Targa features a transverse V-twin engine, shaft drive, twin shock absorber rear suspension, and front and rear disc brakes.

BELOW: The 1991 Moto Guzzi 750 is a smaller capacity motorcycle from the Italian manufacturer, a fact that is evident from the machine's designation. The 750 refers to the approximate metric displacement.

SPECIFICATION
Country of origin: ITALY
Capacity: 743.9cc (45.37cu. in.)
Engine cycle: 4-stroke
Number of cylinders: 2
Top speed: 115mph (185kph)
Power: 48bhp @ 6600rpm
Transmission: 5-speed
Frame: Duplex tubular cradle

1991 MOTO GUZZI DAYTONA 1000 I. E.

The sporting Moto Guzzi of the early 1990s was the 1000 Daytona I. E. with fuel injection in place of conventional carburetors. This superbike was produced in limited numbers; Moto Guzzi produced less than 6000 motorcycles in 1992 compared to

ABOVE: The 1000 Daytona features electronic fuel injection manufactured by Weber Marelli. The engine is a double overhead camshaft unit housed in a Koni monoshock frame and with Brembo disc brakes on alloy wheels.

Honda's 3.8 million. Mass production is beyond the resources of smaller companies such as Moto Guzzi.

SPECIFICATION
Country of origin: ITALY
Capacity: 992cc (60.51cu. in.)
Engine cycle: 4-stroke
Number of cylinders: 2
Top speed: 150mph (241kph)
Power: 92bhp @ 7600rpm
Transmission: 5-speed
Frame: Box section single beam

1991 MOTO GUZZI SP 1000 III SPADA

The Spada is a touring motorcycle designed around the same basic transverse V-twin engine but in a milder state of tune. It fea-

BELOW: The 1991 SP 1000 III Spada was a serious touring motorcycle. It features an integrated fairing and a pair of large capacity panniers for this purpose. Underneath this is the 948.8cc (57.87cu. in.) transverse V-twin engine and shaft drive typical of the company.

tures shaft drive, an integrated fairing and panniers, which all help in its touring role.

SPECIFICATION
Country of origin: ITALY
Capacity: 948.8cc (57.87cu. in.)
Engine cycle: 4-stroke
Number of cylinders: 2
Top speed: 121mph (195kph)
Power: 71bhp @ 6800rpm
Transmission: 5-speed
Frame: Duplex tubular cradle

1991 MOTO GUZZI 1000 CALIFORNIA III

The name of this motorcycle indicates where Moto Guzzi saw its market. The California III is one of several machines that have something of the factory custom in them and which are also suited to touring. Another range of similar bikes has been

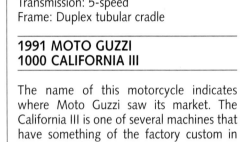

BELOW: The 1994 Moto Guzzi California was one of several models to be so named. A factory custom with pullback bars and dual seat, the screen and panniers are removable.

of the first Moto Morini machines to become widely available outside Italy. The 344cc (20.98cu. in.) 72° V-twin engine powered the machine in both Strada and Sport versions but the Sport develops four extra horsepower and has a riding position similar to that found on a sports bike.

SPECIFICATION
Country of origin: ITALY
Capacity: 344cc (20.98cu. in.)
Engine cycle: 4-stroke
Number of cylinders: 2
Top speed: 96mph (155kph)
Power: 38bhp @ 8200rpm
Transmission: 6-speed
Frame: Tubular cradle

1991 MOTO MORINI DART 350

This is a medium capacity sports bike that features enclosed bodywork and a wrap-around front mudguard. The V-twin engine is completely hidden behind the bodywork, suspended from an aluminum beam frame.

SPECIFICATION
Country of origin: ITALY
Capacity: 344cc (20.98cu. in.)
Engine cycle: 4-stroke
Number of cylinders: 2
Top speed: 105.6mph (170kph)
Power: 34bhp @ 8500rpm
Transmission: 6-speed
Frame: Aluminum beam

BELOW: The Moto Morini 350 Dart of 1991 featured almost completely enclosed bodywork including the front mudguard under which is an aluminum beam frame. Beneath this is mounted an overhead-valve, four-stroke V-twin engine and a six-speed transmission.

produced for police use which is an important market for the company. For many years Moto Guzzi kept a single-cylinder machine in production purely in order to fill such contracts.

SPECIFICATION
Country of origin: ITALY
Capacity: 948.8cc (57.87cu. in.)
Engine cycle: 4-stroke
Number of cylinders: 2
Top speed: 118mph (190kph)
Power: 67bhp @ 6800rpm
Transmission: 5-speed
Frame: Duplex tubular cradle

MOTO MORINI

Alphonso Morini was a motorcycle road-racer who competed on motorcycles of his own design and manufacture. In 1937 he opened a factory to produce motorcycles and commercial three-wheeler machines. The factory was destroyed by Allied bombing during 1943 but in the postwar years a new factory was completed. Morini manufactured 125 and 175cc (7.62 and 10.67cu. in.) four-strokes and had considerable success in racing events. One of these race motorcycles was the double overhead camshaft Rebello which had a displacement of 173cc (10.55cu. in.). Alphonso Morini died in 1969 and the management of the company was taken over by his daughter, Gabriella.

1957 MOTO MORINI REBELLO

The two-stroke Rebello was built for racing in two different displacements: 173 and 246cc (10.55 and 15cu. in.), each with double overhead camshaft engines. Tarquino Provini, who was Morini's leading race rider, achieved much success on the Rebello.

ABOVE: The Moto Morini factory built some motorcycles especially for racing. One such was the double overhead camshaft Rebello of 1957. A very successful design, it brought much prestige to the company.

SPECIFICATION
Country of origin: ITALY
Capacity: 173cc (10.55cu. in.)
Engine cycle: 2-stroke
Number of cylinders: 1
Top speed: 120mph (193kph)
Power: 22bhp @ 9000rpm
Transmission: 4-speed
Frame: Tubular cradle

1977 MOTO MORINI STRADA

Under the control of Morini's daughter the Strada was introduced in 1971 and was one

MÜNCH

This company was named after its founder, Friedl Münch, and he built his first machine in 1965. It was a handcrafted motorcycle and it was only produced in small numbers to special order.

1977 MÜNCH 4 1200 TTS

This machine was technically advanced for its day and its design utilized expensive components and materials such as magnesium alloy to minimize its not inconsiderable weight. The model's engine is an air-cooled, transverse-mounted, NSU four-cylinder car engine and makes the machine suitable for high speed touring. It was updated and fitted with a 1278cc (77.95cu. in.) engine also from NSU.

SPECIFICATION
Country of origin: GERMANY
Capacity: 1177cc (71.79cu. in.)
Engine cycle: 4-stroke
Number of cylinders: 4
Top speed: 124mph (200kph)
Power: 95bhp @ 6000rpm
Transmission: 4-speed
Frame: Tubular steel

BELOW: The Münch Mammoth motorcycle was based around an NSU car engine. Machines of varying capacities including 1200 and 1600cc (73.20 and 97.60cu. in.) were built in limited numbers and the company owner also worked to produce a redesigned Indian Scout.

MUSTANG

An American company – Gladden Products – who had manufactured aircraft components throughout World War II made small capacity motorcycles for basic transport in the USA between 1946 and 1964. Their simple machine initially featured a Villiers engine but this was replaced at a later date by a 314cc (19.15cu. in.) single-cylinder, side-valve engine that had originally been designed as an industrial unit. The machine

BELOW: The Mustang, such as this 1953 model, was like a Cushman in that it was a small machine on small wheels with some of the styling of a larger American motorcycle.

was something of a hybrid of a scooter and a motorcycle. It looked like a motorcycle but was fitted with 12in. (300mm) diameter wheels, for example.

MV AGUSTA

This aeroplane-making company started in the motorcycle business in 1946 and its factory at Cascina Costa, Italy, has produced a huge number of legendary sports bikes. The company was owned by the Agusta family and also made helicopters. Its racing bikes, ridden by some of motorcycle racing's legendary names such as Surtees, Agostini and Hailwood, have won over 40 world championships. The company started helicopter

production which ultimately subsidized motorcycle manufacture. This led to the motorcycle division of the business being absorbed by EFIM, an Italian Government finance group that also controlled Ducati, and which merged the two companies. Later it was acquired by the Cagiva group.

1977 MV AGUSTA 750S AMERICA

This machine is truly one of the most exotic motorcycles of all time. It is an exclusive sports bike designed, as the name implies, primarily for the North American market. The four-cylinder engine offered high performance (described as 'scintillating' at the time of its introduction) and sports bike handling which was achieved through use of Ceriani suspension.

SPECIFICATION
Country of origin: ITALY
Capacity: 790cc (48.19cu. in.)
Engine cycle: 4-stroke
Number of cylinders: 4
Top speed: 122mph (196kph)
Power: 75bhp @ 8500rpm
Transmission: 5-speed
Frame: Duplex tube cradle

BELOW: The MV Agusta 750S America is one of the legendary machines from the 1970s. Built for the American market, it was fitted with a four-stroke, four-cylinder engine.

MZ

As the Iron Curtain descended across Germany in the years immediately after World War II, the MZ factory was created from IFA which had been established in the DKW factory at Zschopau in what became East Germany. MZ – Motorradwerk Zschopau – is the brandname of the IFA Kombinat-Zweiräder group, the nationalized East German motorcycle manufacturers. The company became a prolific manufacturer of two-stroke machines and had considerable road racing success against a background of Cold War politics, racers' defections and East German shortages. This success was due entirely to the talents of Walter Kaaden and a small group of riders. One of their greatest moments of glory was Mike Hailwood's win in the 1963 250 GP at Sachsenring, Germany. MZ also provided the bikes for the DDR entry in the ISDT.

With the reunification of Germany it was possible that MZ would be forced to cease motorcycle production as they now faced stiff competition from modern Japanese and European machines without the aid of DDR state subsidy. However, the company quickly shifted its emphasis to bring modern competitive machines into production.

BELOW: MZ, the impoverished East German motorcycle manufacturer, built two-stroke race machines such as this 1962 model but were able to get the greats like Mike Hailwood to ride for them in Grands Prix. Hailwood rode an MZ at both Monza and Sachsenring.

ABOVE: The MV Agusta 750 America was one of a small number of motorcycles built by the Italian, Count Domenico Agusta, whose company also manufactured helicopters. This motorcycle is regarded as truly exotic.

1956 MZ ES250

The ES250 was a strictly utilitarian motorcycle which was introduced in 1956. It was based around a 243cc (14.82cu. in.), two-stroke, single-cylinder engine and featured Earles' leading link forks and a headlamp mounted in a large nacelle. The machine gained a reputation for its reliability and the cheapness of both its purchase and operation. As such, it had a lengthy production run up until 1973.

SPECIFICATION
Country of origin: DDR
Capacity: 243cc (14.82cu. in.)
Engine cycle: 2-stroke
number of cylinders: 1
Top speed: 72mph (117kph)
Power: 17.5bhp
Transmission: 3-speed
Frame: Tubular cradle

LEFT: The two-stroke MZ ES250 was introduced in 1956 and had a lengthy production run behind the Iron Curtain because the state-run industries were not concerned with fashion. It was both cheap to buy and to run.

BELOW: A later two-stroke but equally basic motorcycle from MZ was the TS250. It too had a long production run that only ended with the collapse of the Iron Curtain and its controlling regime. This is a 1976 model.

1976 MZ TS 250/1

Despite the MZ factory's production of a number of race-winning motorcycles on a shoestring budget, the company's primary concern was the manufacture of the most basic of street bikes intended for use in an austere political and economic climate. The TS 250/1 had a smaller capacity stablemate, the TS 125 – the numerical suffixes relating to the approximate metric displacement of the engines.

SPECIFICATION
Country of origin: DDR
Capacity: 244cc (14.88cu. in.)
Engine cycle: 2-stroke
Number of cylinders: 1
Top speed: 81mph (130kph)
Power: 21bhp @ 5500rpm
Transmission: 5-speed
Frame: Steel tubular

1994 MuZ SKORPION

The company renamed itself MuZ and sought to produce the Skorpion which was designed by British engineers. It features a liquid-cooled, Japanese (Yamaha), single-cylinder engine and a number of Italian components. Wheels are cast from alloy and the bike looks decidedly modern.

SPECIFICATION
Country of origin: GERMANY
Capacity: 660cc (40.26cu. in.)
Engine cycle: 4-stroke
Number of cylinders: 1
Top speed: 108mph (175kph)
Power: 48bhp @ 6250rpm
Transmission: 5-speed
Frame: Twin tube steel

BELOW: The MuZ Skorpion was a modern machine designed for what was intended to be a new MZ company. Sadly, before many could be built the company went out of business.

N

NER-A-CAR

The Ner-a-Car was designed by J. Neracher in America and manufactured from 1921 onward. The motorcycle had a low-level frame constructed from steel channel in a similar way to a car chassis. The machine utilized a 238cc (14.51cu. in.) two-stroke engine that drove the rear wheel through a friction drive. Production was later transferred over to England where certain models were built with 347cc (21.16cu. in.) Blackburne side-valve engines. Some variations made to the Ner-a-Car in its relatively short production run included the fitting of deeply valanced fenders and variations in the front fork design.

ABOVE: A 1923 285cc (17.38cu. in) two-stroke, Simplex-engined Ner-a-car. Other machines were powered by Blackburne side- and overhead-valve engines.

NEW HUDSON

This English company was founded in 1903 and manufactured a variety of machines with a number of different engines, mostly of medium size displacement in both side-valve and overhead-valve configurations. During the 1920s the company had some success in speed records that were set at Brooklands, England, with Bert Le Vack riding the New Hudson. One of their greatest moments came in the 1927 Senior TT in The Isle of Man where Jimmy Guthrie took second place in the race. By the 1930s the company had a range of singles in production. They were available in various displacements and in both side-valve and overhead-valve configurations in each capacity. As a result the Model 80 was a side-valve 250cc (15.25cu. in.) while the Model 91 was the overhead-valve model. It was the same in the 346cc (21.10cu. in.) class where there were models 83 and 85 and again in the 496cc (30.25cu. in.) class with models 84 and 86.

In 1931 the range was redesigned and many New Hudsons were partially enclosed with panels that ran down the sides of the engine – an idea attributed to a motorcycle salesman, Vic Mole. However, It was not particularly successful. By way of a promo-

ABOVE: A 1927 New Hudson of overhead-valve configuration and 346cc (21.10cu. in.) displacement. In the same year Jimmy Guthrie rode a similar machine to second place in the 1927 Senior TT on the Isle of Man.

tional gimmick the company equipped a 548cc (33.42cu. in.) model with a large Watsonian sidecar and rode it from Brooklands to Land's End, Cornwall, England, and back ten times to prove their machine's worth. For their efforts they earned a certificate from the ACU. The company was relying on this publicity to increase the public's enthusiasm for their enclosed models which had not proved as successful sales-wise as they had hoped. The company revamped its range for 1932 and continued with the range for 1933 but motorcycle sales were depressed because of the economic climate and the company turned to the manufacture of brakes parts for Girling. Much later BSA briefly revived the marque as the name for an autocycle with a 98cc (5.97cu. in.) Villiers engine.

NEW IMPERIAL

The company was founded in England in 1892 but it was not until 1910 that successful production motorcycles were built using JAP engines. During the 1920s the firm had a number of Isle of Man TT victories in the 250 Class. In 1924 they won the 250 and 350 classes. The company produced its own engines from 1926 onward and through the 1930s built innovative machines. For 1930 the company offered six models, all based around vertical engines, magneto ignition and three-speed gearboxes. Variations were in valve configuration and displacement.

The same models continued into 1931 but the range was changed for 1932 as the company moved toward unit construction of engines and gearboxes. These updated motorcycles featured wet sump lubrication, twin port heads, inclined cylinders and were available in both 344 and 499cc (20.98 and 30.43cu. in.) displacements. In actual fact, numerous variations of these models were manufactured. A special V-twin of 491cc (29.95cu. in.) was manufactured for racing in 1934. It was forced to retire from the Isle of Man TT but did win The Motorcycle Cup for being the first-ever British-made multi-cylindered motorcycle to cover more than 100 miles (161km) in one hour on British soil. Ginger Wood covered in excess of 102 miles (164km) in the hour at Brooklands to take the cup for New Imperial. New Imperial's last Isle of Man TT win came in 1936, the same year as they won the Ulster Grand Prix in Ireland.

The company began to experience financial problems in the run-up to World War II and these difficulties were compounded by the untimely death of one of the company's founders, Norman Downs, in 1938. Jack Sangster, who was boss of Triumph and Ariel, bought New Imperial intending to move it from the English city of Birmingham to Coventry. World War II intervened and the marque was not revived afterward.

BELOW: A 246cc (15cu. in.) New Imperial racing motorcycle from the late 1930s. By this time the company was in financial trouble and it was bought by Jack Sangster in 1938. World War II saw the company cease motorcycle production and it was not resumed in the postwar years.

ABOVE: A telescopic-forked 1955 Nimbus overhead camshaft, four-cylinder machine of 746cc (45.50cu. in.) displacement. It used a pressed steel frame. Despite upgrades, Nimbus products remained very similar over 37 years.

NIMBUS

This Danish company started in business in 1920 and eventually became Denmark's largest motorcycle factory before stopping production in 1957. In the period it was in business the company, run by Fisker and Nielsen, concentrated on a single model. This was a 746cc (45.50cu. in.) in-line four-cylinder machine. The engine was upgraded from an inlet-over-exhaust valve configuration to an overhead-valve design during the production run. The frame was made from pressed steel and the forks were originally of a trailing arm design but this was eventually superseded by a telescopic type. Despite these upgrades and changes to the shape of the tank and fenders, the Nimbus machines are remarkably similar to each other even when manufactured more than 30 years apart.

NORMAN

This British company based in Ashford, Kent, started in the motorcycle business with the production of 98cc (5.97 cu. in.) autocycles and 122cc (7.44cu. in.) motobykes in the late 1930s. Both machines were fitted with Villiers two-stroke engines. In the postwar years similar machines comprised the range until 1948. In this year the machines were all updated. The autocycle received a 99cc (6.03cu. in.) 2F Villiers engine while the other models now used the 122cc (7.44cu. in.) 10D and 197cc (12.01cu. in.) 6E engines in a rigid frame with telescopic forks.

There were no further changes until 1953 when economy versions and swing arm frames were introduced. A 197cc (12.01cu. in.) trials version was introduced too. A range of mopeds were introduced during the 1950s with names like Nippy and Lido and so the company stayed in production until the end of the decade. The firm was acquired by the British cycle manufacturer Raleigh in 1961 and motorcycle production was dropped.

1956 NORMAN NIPPY

New for 1956 was the Norman Nippy moped that was aimed at the commuter and shopper market as basic and cheap transportation. The machine featured leading link forks and a small Sachs-made two-stroke engine. Over the longer term, it was intended that Norman would increase production of its own components and gradually reduce what needed to be bought in. Furthermore, it was hoped that the moped would eventually replace the company's autocycle models.

SPECIFICATION
Country of origin: GREAT BRITAIN
Capacity: 47.6cc (2.90cu. in.)
Engine cycle: 2-stroke
Number of cylinders: 1
Top speed: n/a
Power: n/a
Transmission: 2-speed
Frame: Pressed steel beam

NORTON

This is one of the legendary names from the British motorcycle industry. The company was founded by James Lansdowne Norton in 1901. His first motorcycles used Swiss Moto–Reve and French Peugeot engines and were built in England under license. Aboard a V-twin Norton, H. Fowler won the twin-cylinder class at the 1907 Isle of Man TT. A single of 633cc (38.61cu. in.) was marketed in 1908 and another model of 490cc (29.89cu. in.) in 1911. Norton produced their first overhead-valve single in 1922. In 1924 Nortons won both the Sidecar TT and the Senior TT on the Isle of Man with riders and passenger being Tucker, Moore and Bennett respectively.

In 1925 James Landsowne Norton passed away but the company continued in business. Sidecar passenger Walter William Moore was responsible for designing the first overhead camshaft Norton before leaving to work for NSU in Germany. Some felt that Norton's roadgoing bikes looked dated by the end of the 1920s and a revamp was not long in coming. Arthur Carroll replaced Moore in the design department and redesigned Norton's engine into something that would endure until 1963. Norton models were given numerical designations and in the 1930s consisted of machines such as the 348cc (21.22cu. in.) JE and CJ, the 490cc (29.89cu. in.) 16H, and the 588cc (35.86cu. in.) Model 19 amongst others.

Norton introduced a four-speed gearbox in 1933. The range was gradually updated through the 1930s. The last years before World War II saw the introduction of the Racing International models (also referred to as the Manx Grand Prix). These motorcycles were the foundations on which the postwar Manx Norton was based. Through the war years Norton's Bracebridge Street works supplied 100,000 of their 16H models to the British Army. In smaller numbers they supplied the Big 4 sidecar outfit with a driven sidecar wheel to make it suitable for cross-country use.

The 16H was reintroduced after the war in a slightly updated frame but the major postwar news was the introduction of the Manx Norton. It was introduced for TT racing, hence its 'Manx' tag, but gradually became more and more developed. The famous featherbed frame was introduced as the basis for the Manx but also used in the Models 88 and 99. Norton was acquired by Associated Motorcycles Ltd. in 1956. During the 1960s they experimented with partially enclosed motorcycles, notably in the 250cc (15.25cu. in.) Norton Jubilee.

The firm moved from Birmingham to London, England, in 1963 and in the following years Norton motorcycles became more of a combination of Norton, AJS and

BELOW: In the years after World War II Norton reintroduced its prewar designs but gradually upgraded them to models such as the 18 and ES2. This machine featured telescopic forks but still used a rigid frame and side-valve engine.

Matchless parts. In 1966 AMC went into liquidation and ceased to exist. Dennis Poore of Manganese Bronze bought the company and transferred production of the 647 and 746cc (39.50 and 45.50cu. in) twins for the US market, especially the Commando, to the former Villiers factory in Wolverhampton. A little later in 1970 he became Chairman of Norton–Villiers Ltd. What followed were the death throes of the British bike industry as Norton–Villiers absorbed BSA and Triumph to become Norton–Villiers–Triumph and subsequently NVT Motorcycles Ltd. In these latter years , the company persevered with the Wankel rotary-engined Norton motorcycle.

1942 NORTON MODEL 16H

The Norton 16H was a side-valve, single-cylinder machine in production long prior to World War II. The 490cc (29.89cu. in.) sin-

ABOVE: The Norton 16H was a 490cc (29.89cu. in.) side-valve, single-cylinder machine manufactured in large quantities for the British and Allied armies of World War II. The motorcycle featured a rigid frame and girder forks.

gle was initially fitted with a three-speed gearbox but upgraded to a four-speed in 1935. The 16H was seen as the most suitable for military purposes during the months leading up to to the outbreak of war and it remained in production as Norton's military motorcycle. It was reintroduced in civilian colors after the war.

SPECIFICATION
Country of origin: GREAT BRITAIN
Capacity: 490cc (29.89cu. in.)
Engine cycle: 4-stroke
Number of cylinders: 1
Top speed: 60mph (96.5kph)
Power: n/a
Transmission: 4-speed
Frame: Tubular steel

1953 NORTON DOMINATOR 77

A new frame appeared in 1953 from the Bracebridge Street works and, like the famous featherbed, it featured a pivoted rear fork but was based on a cradle design. It was used as the basis for the Model 7 Dominator and the ES2. A Model 88 Dominator using the featherbed frame was also made in 1953 for export only. The machine was powered by a four-stroke

BELOW: The Dominator was introduced as the Model 7, a twin, in 1949 with an engine design by Bert Hopwood. It was a success and ensured the Dominator name would remain in use for many years thereafter. This is a 1953 Model 77 Dominator.

overhead-valve twin and proved a success to the extent that the Dominator name would remain in use well into the 1960s. The later Dominators utilized a construction that came to be termed 'the slimline frame' as well as incorporating a larger capacity twin engine.

SPECIFICATION
Country of origin: GREAT BRITAIN
Capacity: 497cc (30.31cu. in.)
Engine cycle: 4-stroke
Number of cylinders: 2
Top speed: n/a
Power: n/a
Transmission: 4-speed
Frame: Tubular steel

stroke engine, making it a very formidable machine.

SPECIFICATION
Country of origin: GREAT BRITAIN
Capacity: 498cc (30.37cu. in.)
Engine cycle: 4-stroke
Number of cylinders: 1
Top speed: 140mph (225kph)
Power: 47bhp @ 6500rpm
Transmission: 4-speed
Frame: Double loop cradle

1971 NORTON COMMANDO FASTBACK

The Fastback was a café racer styled version of the Commando launched in 1969. It featured what were termed 'Isolastic' rubber mounts for the engine, gearbox, swingarm, exhaust system and rear wheel. These components were bolted together and attached to the frame by the Isolastic mounts. Roadholder forks were fitted with a hub featuring a twin leading-shoe brake.

SPECIFICATION
Country of origin: GREAT BRITAIN
Capacity: 745cc (45.44cu. in.)
Engine cycle: 4-stroke
Number of cylinders: 2
Top speed: 115mph (185kph)
Power: 56bhp @ 6500rpm
Transmission: 4-speed
Frame: Tubular steel

BELOW: A 1971 Norton Commando Fastback. This café racer style of machine featured a rubber-mounted engine to minimize vibration and the renowned Roadholder forks.

1960 MANX NORTON

Norton had a long history of building overhead camshaft racing singles. There were updates and improvements and even models with different capacities, such as the short-stroke design introduced in 1954, and also the 350cc and 500cc class engines. The famous 'feather-bed' frame, which had been developed by Rex McCandless, was made commercially available during the early 1950s after Norton's 1-2-3 wins in both the Junior and Senior TTs at The Isle of

ABOVE: The Manx Norton became a legend as a roadracing overhead camshaft single. It used the famous McCandless featherbed frame and by 1960 was a potent race bike. Its name reflects its Isle of Man racing success including a 1-2-3 win in the Senior TT of 1950.

Man in 1950. By 1960 the Manx Norton was approaching the peak of its development as a racing machine. It featured the featherbed frame, an AMC gearbox (as a result of Norton's acquisition by AMC), Roadholder telescopic forks and the short-

SPECIFICATION
Country of origin: GREAT BRITAIN
Capacity: 828cc (50.50cu. in.)
Engine cycle: 4-stroke
Number of cylinders: 2
Top speed: 115mph (185kph)
Power: 58bhp @ 5900rpm
Transmission: 4-speed
Frame: Duplex tubular cradle

1990 NORTON COMMANDER

The rotary project was ongoing for Norton for over a decade and the Commander was one motorcycle to use the rotary in a modern and touring guise. At its heart is the twin-chamber, liquid-cooled rotary engine hidden behind the angular fairing and hard luggage. In other respects the Commander was typical of touring motorcycles of the 1990s with alloy wheels and twin discs at the front. Such machines as this found favor with police departments and some other emergency services.

Norton developed a race version of the rotary-engined machine, the NRS588. It was successfully raced on several occasions, including Steve Hislop's 1992 Isle of Man Senior TT victory, which reminded the company of past glories on that famous circuit.

SPECIFICATION
Country of origin: GREAT BRITAIN
Capacity: 588cc (35.86cu. in.)
Engine cycle: Rotary
Number of cylinders: Not applicable
Top speed: 124mph (200kph)
Power: 85bhp @ 9000rpm
Transmission: 5-speed
Frame: Pressed steel monocoque

1977 COMMANDO MARK III

The Norton Commando was introduced in the late 1960s as a fast and refined version of the parallel twin. The early models were of 750cc (45.75cu. in.) displacement and reached almost 120mph (193kph). In 1973 the 828cc (50.50cu. in.) models were introduced as sports bikes and were built with a rubber-mounted engine intended to reduce the vibration felt by the rider. However, the Mark III was introduced in 1976 with electric start and disc brakes. It appealed to the US market and could compete with the Japanese in terms of refinement but financial problems at the manufacturers meant it was soon dropped in favor of the Triumph Trident which was perceived to be a more modern machine.

ABOVE: The last of the 850 Norton Commandos were made in 1977 while the Wolverhampton, England, factory was in the hands of the Official Receiver. Electric starter and disc brakes front and rear were two of the upgrades aimed at competing with Japanese makers.

BELOW: The Norton Commando was seen as one of the machines that would save the British bike industry. Norton twin production was shifted to the Villiers factory to become Norton–Villiers Ltd. This company later became Norton–Villiers–Triumph (NVT).

LEFT: The Norton Commander was a rotary-engined motorcycle. Rotary engines never seem to have proved particularly popular but Norton persevered with just such an engine and produced a workable design. The engine was used in several Norton models.

ABOVE: The Norton Commando was introduced
in the 1960s as a refined version of the
traditional British parallel twin. In the final
750cc form, shown here, the motorcycle was
capable of 120mph (193kph).

1992 NORTON F1 SPORT

The F1 was a sports version of the rotary-engined Norton motorcycle. It was based around the same basic engine, albeit one tuned for a higher performance. The fairing is more sporting with an overall lower profile and components such as the brakes are to a higher specification.

SPECIFICATION
Country of origin: GREAT BRITAIN
Capacity: 588cc (35.86cu. in.)
Engine cycle: Rotary
Number of cylinders: Not applicable
Top speed: 140mph (225kph)
Power: 95bhp @ 9500rpm
Transmission: 5-speed
Frame: Twin spar alloy beam

BELOW: The 1992 Norton F1 Sport was a roadgoing version of the NRS588 race bike. Like the other machines it relied on the Wankel rotary engine in which a three-sided rotor turns in an eccentrically-shaped chamber.

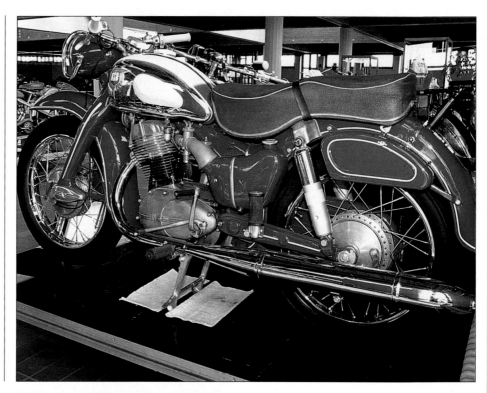

ABOVE: An NSU Supermax 248cc (15.12cu. in.) double overhead camshaft, twin-cylinder machine of the 1950s, designed by Albert Roder and developed by Ewald Praxl and Walter Froede. A smaller model, the 174cc (10.61cu. in.) Maxi, was also built using an overhead camshaft, single-cylinder engine.

NSU

NSU stands for Neckarsulm Strickmaschinen Union, a German company which began making knitting machines and bicycles around the turn of the 19th century. By 1903 the company was manufacturing its own motorcycles after experimenting with a bicycle-style frame and a Swiss Zedel engine. Its own machines featured a 329cc (20.06cu. in.) engine. These were followed by a V-twin engine of 804cc (49.04cu. in.) displacement. NSU went on to make a variety of machines in the 1920s, including a number of racing bikes.

The company also employed Walter William Moore, formerly with Norton, to design some new single-cylinder bikes in the 1930s.

In the last years before World War II the company produced supercharged vertical twins for racing. Teething troubles stopped them winning and then the war intervened. The NSU factory made equipment for the German Army during the war and returned to civilian production in 1949. One of the first new models was the 98cc (5.97cu. in.) Fox with a pressed steel frame, leading link forks and cantilever rear suspension. Car manufacture took precedence over motorcycle production and two-wheelers were last made by NSU in 1965.

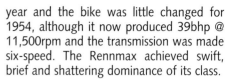

1942 NSU KETTENKRAD

This half-track machine is one of motorcycling's true oddities. While three-wheeler variants of motorcycles, sidecars, forecars and trikes have been relatively common through motorcycling's history, the half-tracked machine has not. The Kettenkrad was built as a light gun tractor and personnel transporter for use in the difficult conditions experienced in war such as snow and sand, in the Soviet Union and North Africa respectively. It used a number of existing motorcycle and automotive components such as an Opel truck engine and specially made parts.

SPECIFICATION
Country of origin: GERMANY
Capacity: 1478cc (90.15cu. in.)
Engine cycle: 4-stroke
Number of cylinders: 4
Top speed: 45mph (72kph)
Power: n/a
Transmission: 6-speed
Frame: Welded steel hull

ABOVE: The NSU Kettenkrad was one of World War II's oddest vehicles and was essentially a half-track motorcycle. It was designed to cope with the terrible conditions of the Eastern Front and the deserts of North Africa. Opel manufactured the four-cylinder engine.

1952 NSU RENNMAX

The overhead camshaft Rennmax was developed by Dr Walter Froede out of a scrapped project. He used a sports twin crankcase and a pair of barrels from a tour. The bike soon proved dominant on the race-track in the 250 class. It took the top four places in the Lightweight 1952 Isle of Man TT. The engine was developed and the first Rennmax delivered 27bhp @ 9000rpm. Later in that same year, after further adjustment, the machine could deliver 31bhp @ 10,400rpm. Claiming second place in the Italian Grand Prix, between two Moto Guzzis, assured its fame.

The frame was changed for a pressed steel beam for 1953 and a fairing was fitted. Racing success continued through the

year and the bike was little changed for 1954, although it now produced 39bhp @ 11,500rpm and the transmission was made six-speed. The Rennmax achieved swift, brief and shattering dominance of its class.

SPECIFICATION
Country of origin: GERMANY
Capacity: 250cc (15.25cu. in.)
Engine cycle: 4-stroke
Number of cylinders: 2
Top speed: n/a
Power: 31bhp @ 10,400rpm
Transmission: 4-speed
Frame: Duplex cradle

1972 NSU QUICKLY S

This was a moped of the step-through design intended as cheap and basic transportation, primarily for use in urban areas. It was first produced in 1953 and enjoyed a long production run. The engine was made from alloy and the frame from pressed steel. Final drive was by chain from the two-speed gearbox and a multiplate clutch.

SPECIFICATION
Country of origin: GERMANY
Capacity: 49cc (2.98cu. in.)
Engine cycle: 2-stroke
Number of cylinders: 1
Top speed: 30mph (48kph)
Power: 1.39bhp
Transmission: 2-speed
Frame: Pressed steel beam

NUT

NUT is an acronym that stands for Newcastle Upon Tyne and was the name given to motorcycles produced in that English city spasmodically in the first decades of the 20th century. The company was founded in 1911 and a rider aboard a NUT machine won the Junior TT in the Isle of Man in 1913. It was financial problems at the company that ensured NUT motorcycle production was not continuous. A range of models was introduced in 1931 using JAP V-twins in 500 and 700cc (30.50 and 42.70cu. in.) overhead-valve configurations as well as a 750cc (45.75cu. in.) side-valve. The machines were conventional in design and were manufactured to a high standard. The range was continued and extended for 1932 and 1933. However, production was minimal and finally ceased permanently in the latter year.

LEFT: NUT used various V-twins during their 20 year period of manufacture, including the side-valve JAP 976cc (59.53cu. in.) version fitted to this 1922 Model V. This motorcycle also features a diamond frame, girder forks and a three-speed gearbox.

O

OD

This was a fairly short-lived German company run by Wily Ostner that produced motorcycles between 1927 and 1935 in a Dresden factory. The company's initials stood for Ostner Dresden. The motorcycles were finished in gray-blue and reputed to be of good quality. The company produced a range of motorcycles from 347cc (21.16cu. in.) to 996cc (60.75cu. in.) mostly using MAG proprietary engines although a number of racing motorcycles and sidecar outfits used JAP engines. OD were the only German manufacturer of the late 1920s and early 1930s to offer a reverse gear. OD also produced aluminum framed two-stroke motorcycles of 198 and 246cc (12.07 and 15cu. in.) displacement. In its last years, the company concentrated on the manufacture of three-wheelers, made both before and after World War II.

1931 OD TS50

This machine was typical of OD's motorcycles, being constructed from a range of proprietary components including an inlet-over-exhaust valve configuration MAG single, a Hurth gearbox and a Bosch magneto.

SPECIFICATION
Country of origin: GERMANY
Capacity: 498cc (30.37cu. in.)
Engine cycle: 4-stroke
Number of cylinders: 1
Top speed: 70mph (112kph)
Power: 13bhp @ 3500rpm
Transmission: 3-speed
Frame: Tubular cradle

OEC

OEC stands for Osborn Engineering Company and the company was founded by Frank Osborn who built his first motorcycles in the early years of the 20th century using Minerva and MMC engines. This early production ceased but was resumed by Osborn's son, John, in 1920. The OEC factory was based in Gosport, Hampshire, England. The company went on to become more involved with record breaking in the 1920s and 1930s.

Fred Wood developed the duplex steering on OEC machines from the late 1920s when it was fitted to the company's duplex cradle frame. It used a complex set of links and pivots which gave a high degree of

BELOW: The Osborn Engineering Company manufactured the Commander in 1938 as one of range of models whose names had nautical connections. Each model used an AJS-manufactured engine prior to World War II.

self-centering to the steering. In 1930 duplex steering was standard on OEC models although rear suspension was an option. The company used Blackburne, JAP and Sturmey Archer engines of displacements ranging from 350 to 500cc (21.35 to 30.50cu. in.) in both side- and overhead-valve configurations. The variety continued through the decade although engines and suppliers varied. OEC also manufactured 498 to 1000cc (30.37 to 61cu. in.) V-twin motorcycles in a similarly varying way. In 1934 Matchless V-twins and Villiers singles were listed.

In 1936 the company introduced the innovative Atlanta Duo. This machine featured duplex steering, a feet-forward riding position and partial enclosure. By 1937 OEC were using AJS engines and in 1938 had a range of motorcycles that were named Commander, Cadet, Commodore and Ensign. Production stopped for the duration of World War II and in 1949, when the company restarted in business, OEC machines were totally conventional with Villiers 10D and 6E engines in rigid frames equipped with telescopic forks. In this guise OEC production continued until 1954.

1934 OEC ATLANTA DUO

This was an unusual design of motorcycle which incorporated a seat height of only 21in. (530mm) achieved through the use of

ABOVE: The Atlanta Duo was an unusual machine with a feet-forward riding position, footboards and OEC's suspension.

a tubular frame that at the rear carried plunger suspension and at the front OEC's duplex steering system. The mechanicals were fitted into the frame and there was a choice of three different engines: overhead-valve, singles or a V-twin. The dual seat sat atop the tubes and had a cowl ahead of it, a wide rear mudguard and a back rest. Footboards ran down each side of the machine and up into legshields at the front.

SPECIFICATION
Country of origin: GREAT BRITAIN
Capacity: 750cc (45.75cu. in.)
Engine cycle: 4-stroke
Number of cylinders: 2
Top speed: n/a
Power: n/a
Transmission: 4-speed
Frame: Tubular steel

OK SUPREME

OK were founded in 1899 and their early motorcycles used De Dion, Minerva, Precision and Green engines. In the interwar years the company manufactured motorcycles with proprietary engines from

BELOW: This racing machine, built by OK Supreme in 1934, was based on an overhead camshaft, single-cylinder engine of 346cc (21.10cu. in.) displacement. The roadgoing versions were known as the Silver Cloud.

Blackburne, Bradshaw and JAP. In 1928 OK won an Isle of Man TT. The partners who had formed the company split up and one of them, Ernest Humphries, appended Supreme to the marque's name and went on producing motorcycles.

During the 1930s the company only produced four-strokes using engines from JAP and Matchless. In 1932 the company catalogued eight different models that ranged in displacement from 148cc (9.02cu. in.) upward. The firm continued in a similar way until the outbreak of World War II, producing a range of models with something for everyone. Each year OK Supreme produced a slightly adjusted range of models of both side-valve and overhead-valve types. By 1938 there were a total of 14 models in the range. World War II saw OK Supreme's factory transferred to the production of commodities other than motorcycles and full scale production did not resume after the cessation of hostilities.

1932 OK SUPREME LIGHTHOUSE

'Lighthouse' was more of a nickname than one given by the manufacturer but it stuck

ABOVE: The sightglass that earned this 1932 overhead camshaft, 248cc (15.12cu. in.) motorcycle its nickname of 'Lighthouse' is clearly visible at the top of the camshaft drive case under the fuel tank.

because of the design of the model. The inclined engine, a single, had a vertical camshaft drive with a sight glass at the top for camshaft inspection. The machine was an innovative way of producing an overhead camshaft engine and was designed by George Jones.

SPECIFICATION
Country of origin: GREAT BRITAIN
Capacity: 248cc (15.12cu.in.)
Engine cycle: 4-stroke
Number of cylinders: 1
Top speed: 60mph (96.5kph)
Power: 10bhp
Transmission: 4-speed
Frame: Tubular cradle

OPEL

The car manufacturer Opel, now part of General Motors, at various times in its early history produced motorcycles. In the first decade of the 20th century the Russelsheim factory in Germany manufactured motorcycles with single-cylinder engines for some time. Motorcycle production then stopped and was not resumed until after World War I. At this time the company produced a 123cc (7.50cu. in.) bicycle attachment engine which fitted on the left side of a bicycle's rear wheel. From this the company went on to produce a motorcycle of bicycle origin which was equipped with a single-cylinder engine that was fitted within the frame.

During the mid-1920s Opel manufactured 498cc (30.37cu. in.) singles. Later in that same decade they made frames from pressed steel, utilizing 498cc (30.37cu.in.) overhead- and side-valve engines. The production of motorcycles stopped altogether in the early 1930s.

1929 OPEL MOTOCLUB

This was one of the last machines made by Opel and used a design which, at the time, was considered innovative. It consisted of pressed steel forks and frame – a design which had been pioneered on Neander motorcycles. The engine was of Opel's own design and featured overhead-valves and twin exhaust ports.

SPECIFICATION
Country of origin: GERMANY
Capacity: 496cc (30.25cu. in.)
Engine cycle: 4-stroke
Number of cylinders: 1
Top speed: 70mph (112kph)
Power: n/a
Transmission: 4-speed
Frame: Pressed steel

ABOVE: The 1929 Opel Motoclub was innovative for its time, even though it was one of the last motorcycles made by Opel. It featured a pressed steel frame.

OSSA

This company, based in Barcelona, Spain, started its motorcycle production in 1946. Prior to this it had been involved in the production of cinematic projection equipment – its logo shows a projection device. Ossa specialized in off-road motorcycles based around two-stroke engines. Ossa's first motorcycle was a 124cc (7.56cu. in.) model and moped production began in 1951. The worldwide export of Ossa machines started in the 1960s when Ossa introduced the 175 and 230cc (10.67 and 14.03cu. in.) engines that had been designed by Edwardo Giro, son of the founder, Manuel Giro. By 1980 in post-Franco Spain Ossa was receiving government assistance in an attempt to help the company through its financial difficulties. However, by the middle of the 1980s Ossa was forced to cease motorcycle production.

1980 OSSA 500 YANKEE

The 500 Yankee was, unusually for Ossa, a sports road bike. The machine featured cast alloy wheels and disc brakes but was otherwise traditional in its design. It was based around a two-stroke twin engine and the numerical designation approximates to its metric capacity.

ABOVE: The 500 Yankee of 1980 was typical of a sports two-stroke of its era, with alloy wheels and disc brakes. Ossa were better known for building dirt bikes.

SPECIFICATION
Country of origin: SPAIN
Capacity: 488cc (29.76cu. in.)
Engine cycle: 2-stroke
Number of cylinders: 2
Top speed: 115mph (185kph)
Power: 58bhp @ 7500rpm
Transmission: 6-speed
Frame: Duplex cradle

1982 OSSA 350 DESERT

The Desert was available in two displacements of 250 and 302.7cc (15.25 and 18.46cu. in.) and was one of a range of dirt bikes made by the Spanish company. While the Desert is an enduro bike, the company also made a 350 Trial designed for slower speed, off-road competitive use.

SPECIFICATION
Country of origin: SPAIN
Capacity: 302.7cc (18.46cu. in.)
Engine cycle: 2-stroke
Number of cylinders: 1
Top speed: 80mph (128kph)
Power: n/a
Transmission: 5-speed
Frame: Duplex tubular cradle

BELOW: The 1976 Ossa 350 Trial is a two-stroke, single-cylinder trials bike of 302cc (18.46cu. in.) displacement. It features typical technology of the time: tall wheels, drum brakes and swinging arm rear suspension.

P

PANTHER

This company was originally known as P & M after its founders, Jonah Phelon and Richard Moore. The factory was based in West Yorkshire, England. In 1900 the company fitted engines designed by Jonah Phelon to frames made by the Beeston Humber Company. Later Humber went on to build these machines under license. All P & M machines had chain drive and the engine was so designed as to take the place of the front frame down-tube. Typical of the engines were single-cylinder, side-valve and overhead-valve units. The partnership with Moore of 1905 enabled commercial production to begin. Granville Bradshaw redesigned the motorcycle in 1923 and the Panther brandname was adopted although the analogy suggested by this name was not wholly appropriate. The company introduced lightweight singles in 1932 as an economy range of bikes and called them Red Panthers. An earlier line of lightweights were Villiers motorcycles, with two-stroke engines, referred to as Panthettes.

During World War II the company manufactured aircraft parts but resumed motorcycle production after the cessation of hostilities. They stuck to their tried and tested inclined engine formula and produced a range of overhead-valve motorcycles with displacements of 249, 348 and 594cc (15.18, 21.22 and 36.23cu. in. respectively). Telescopic forks appeared in 1947 and production of singles continued until the 1960s. The company also experimented with a range of scooters and mopeds in the 1950s and 1960s.

BELOW: The Panther Cub of the 500cc displacement class. The P & M Company persevered with the sloped forward engine configuration in both side- and overhead-valve forms in their larger capacity motorcycles until the end of production in the 1960s.

1935 PANTHER 100

The name Panther was first used in 1923 when the revised engine was introduced. The engine – then a 500cc (30.50cu. in.) unit – was gradually increased in displacement to 596cc (36.35cu. in.) in 1929 and 650cc (39.65cu. in.) in 1959. The Panther 100 of 1935 was the big single, production of which had continued throughout the Depression years, while a range of smaller capacity machines had also been produced as an economy measure.

SPECIFICATION
Country of origin: GREAT BRITAIN
Capacity: 596cc (36.35cu. in.)
Engine cycle: 4-stroke
Number of cylinders: 1
Top speed: 85mph (140kph)
Power: 26bhp @ 5000rpm
Transmission: 4-speed
Frame: Tubular steel

1964 PANTHER 120S

This was one of the last Panther motorcycles built, as the company was experiencing financial trouble by the early years of

ABOVE: The Panther 120S of 1964 was one of the last models built. By the early 1960s the company was in severe financial trouble and production eventually ceased.

the 1960s. Typical of the company's utilitarian motorcycles, the 120S was a large capacity single with a bore and stroke of 88 and 106mm. It used a Burman gearbox and was a completely traditional machine from the company who had by this time diversified into the production of far more fashionable scooters.

SPECIFICATION
Country of origin: GREAT BRITAIN
Capacity: 645cc (39.34cu. in.)
Engine cycle: 4-stroke
Number of cylinders: 1
Top speed: n/a
Power: 28bhp @ 4500rpm
Transmission: 4-speed
Frame: Tubular steel

PARILLA

Like several other Italian motorcycle manufacturers Parilla's beginnings are in postwar Italy. The firm was named after its founder, Giovanni Parilla, whose first motorcycle was a machine with a 248cc (15.12cu. in.) single-cylinder, overhead camshaft engine. He followed this with a racing version in 1947. From then on the company made a wide variety of singles and twins of up to 350cc (21.35cu. in.). Most were conventional with the exception of the Slughi of 1958. This was an almost entirely enclosed machine that, while a motorcycle, was not dissimilar in appearance to a scooter. A special model of this known as the Impala was made for export to the US market. The factory was acquired by an industrial group which collapsed in 1967, thereby ending motorcycle production completely.

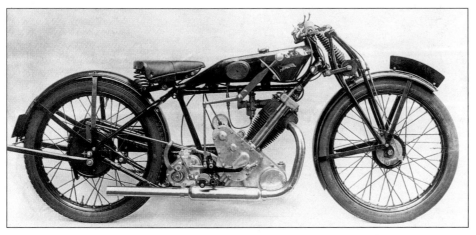

ABOVE: Racing was an aspect of motorcycling taken seriously by Parilla, a company which was founded in Italy in the years following World War II. This is a works racer of the 175cc displacement class.

1958 PARILLA 350 TURISMO

The 348cc (21.22cu. in.) engine, a parallel vertical-twin unit, was used to power the Turismo version of Parilla's late 1950s medium weight motorcycle. It also featured swinging arm rear suspension and telescopic forks. Drum brakes were fitted and final drive was by chain. The motorcycle had a conventional appearance with long mudguards, shrouded forks and a headlamp nacelle.

SPECIFICATION
Country of origin: ITALY
Capacity: 348cc (21.22cu. in.)
Engine cycle: 4-stroke
Number of cylinders: 2
Top speed: 71.5mph (115kph)
Power: n/a
Transmission: 4-speed
Frame: Tubular cradle

PEERLESS

This brandname was used by three different motorcycle companies: two British ones in the years prior to World War I, and an American company. The American company, based in Boston, Massachusetts, produced machines between 1913 and 1916. They used single and V-twin engines of their own design with Bosch Magneto ignition and fitted them into loop frames. Peerless experimented with shaft drive and telescopic forks. The 1912 5hp Peerless was a single-cylinder machine with Bosch magneto, a clutch (described as a free engine idler), a gas tank between the frame rails, a sprung saddle and sprung forks.

PEUGEOT

This French company was founded in France in 1899 and was among the pioneers of the car and motorcycle industries. Peugeot built both singles and twins for other factories including British Norton and Dot. A Peugeot V-twin-engined Norton ridden by H. R. Fowler won the first-ever Isle of Man TT in 1907. Other Peugeot racing bikes were designed by Jean Antoinescu

BELOW: A 1922 Peugeot 500 GP racer. These motorcycles, designed by Jean Antoinescu, were very successful and brought great prestige to the French company during the 1920s. Featuring overhead camshaft engines, they enjoyed race success until 1927.

and these featured gear-driven overhead camshafts and a displacement of 494cc (30.13cu. in.). These machines had not inconsiderable success during the 1920s and were finally retired in 1927.

Peugeot's first successful production model was a 334cc (20.30cu. in.) single. After World War II Peugeot produced V-twins of 295, 344, 738 and 746cc (17.99, 20.98, 45.01 and 45.50cu. in. respectively). They also produced singles of 173, 269 and 346cc (10.51, 16.40 and 21.10cu. in. respectively).

Cycles Peugeot was founded in 1926 to concentrate on motorcycle production. Their model range was redesigned at the end of the 1920s and the new models in various forms continued in production until the outbreak of World War II. In the post-war years the Peugeot operation concentrated on small capacity mopeds and motorcycles following a slump in the sales of larger machines in the mid-1950s.

1980 PEUGEOT TSE

The TSE was a diminutive trail bike produced at the plant in Montbeliard in France. It featured trials-type tires and off-road styling and was an enduro version of the Peugeot TSAL Roadster. Other similar machines made by Peugeot at the time include the SX8T and SX5C.

SPECIFICATION
Country of origin: FRANCE
Capacity: 49cc (2.98cu. in.)
Engine cycle: 2-stroke
Number of cylinders: 1
Top speed: 28mph (45kph)
Power: 2.3bhp @ 5250rpm
Transmission: 2-speed
Frame: Tubular steel

1980 PEUGEOT 103SP

This is a completely orthodox step-through design of moped intended as basic transportation and based around a small and economical two-stroke engine.

SPECIFICATION
Country of origin: FRANCE
Capacity: 49cc (2.98cu. in.)
Engine cycle: 2-stroke
Number of cylinders: 1
Top speed: 28mph (45kph)
Power: 2bhp @ 5000rpm
Transmission: 2-speed
Frame: Step-through

RIGHT: The 103SP was a slightly enhanced version of Peugeot's basic step-through moped, based around a 49cc (2.98cu. in.) engine. At this time, eight differing variations of this model were available.

PIAGGIO

The company was founded by Rinaldo Piaggio in 1884 to manufacture components for ships and diversified into the aeronautical industry in 1915. In the years after World War II Piaggio diversified again into the production of motorized two-wheelers. Piaggio manufactured the famous Vespa range before taking over Gilera in 1969. In the 1990s the company introduced a new range of scooters.

1991 PIAGGIO COSA 200

The Cosa 200 had the appearance of the earlier scooters for which Piaggio was famous but with an updated design. The Cosa 200 uses a two-stroke, rotary-valve, single-cylinder engine which is concealed behind the composite bodywork. The body-

ABOVE: Piaggio can truly claim to have made a considerable contribution to the development of cheap and basic transportation the world over with machines such as this 1977 Vespa 90, one of their four scooters in that year.

work was so designed as to provide storage for the rider's helmet when the machine is parked. Another similar 1991 machine from Piaggio was the PX 50XL Plurimatic scooter of 49.28cc (3cu. in.).

SPECIFICATION
Country of origin: ITALY
Capacity: 198cc (12.07cu. in.)
Engine cycle: 2-stroke
Number of cylinders: 1
Top speed: 61.5mph (99kph)
Power: 11bhp @ 6000rpm
Transmission: 4-speed
Frame: Steel monocoque

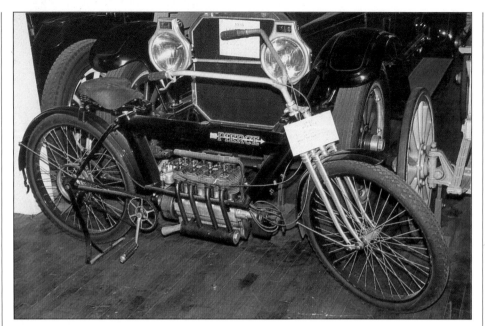

ABOVE: Pope started out manufacturing single-cylinder engined motorcycles but soon moved to V-twin engines as these quickly became the type favored for American bikes.

and the firm later went on to produce a V-twin. The 1914 model displaced 998cc (60.87cu. in.) and had overhead valves.

1913 POPE V-TWIN

The 1913 Pope was in some ways advanced for its time. For example, the frame had a plunger rear suspension set-up and, while there was no front brake, there were two rear brakes. The overhead-valve engine was modern while other aspects were typical of American bikes of the era, with leaf sprung front forks and a three-speed handshift gearbox.

SPECIFICATION
Country of origin: USA
Capacity: 998cc (60.87cu. in.)
Engine cycle: 4-stroke
Number of cylinders: 2
Top speed: 75mph (120kph)
Power: 15bhp @ 3800rpm
Transmission: 3-speed
Frame: Steel loop

PIERCE ARROW

Pierce Arrow were among the pioneers of American motorcycling and also made cars. The company produced a range of motorcycles between 1909 and 1913. These machines included belt-driven singles and an in-line four-cylinder machine which had shaft drive. The four displaced 598cc (36.47cu. in.) The top tube of the frame was of large diameter and carried the fuel.

BELOW: Pierce Arrow motorcycles were advanced for their day with four-cylinder engines and shaft drive, features which only gained wider acceptance in the 1960s.

ABOVE: The Pierce cycle company, makers of the four-cylinder machine shown, were based in Buffalo, New York, USA, and also produced a single-cylinder motorcycle.

A clutch and two-speed gearbox was introduced in 1910.

POPE

This was another of the early American motorcycle manufacturers and named after its founder, Colonel Albert A. Pope. He built cars and bicycles as well as motorcycles between 1911 and 1918. The Pope used a 426cc (26.98cu. in.) single-cylinder engine

PUCH

Puch was founded in Austria and manufactured motorcycles from 1903 onward. During their earliest years they produced a range of singles and V-twins incorporating their own design of engine. After 1923 the major production line was of double piston two-stroke machines designed by Giovanni Marcellino in as many as six capacities. In 1934 Puch amalgamated with Steyr and Daimler to form the grouping known as Steyr–Daimler–Puch. Toward the end of the 1930s the company produced some side-valve flat-fours of 792cc (48.31cu. in.), mainly for military customers.

In the years which followed World War II the Austrian company manufactured a range of two-stroke motorcycles with single piston engines and pressed steel frames. They also updated some of the larger displacement two piston designs such as the 248cc (15.12cu. in.) SGS in a pressed steel frame which stayed in production until the early 1970s. However, by the 1970s, production was concentrated on 49, 123 and 173cc (2.98, 7.50 and 10.55cu. in.) machines including a range of trials and motocross motorcycles, in addition to a line of more basic mopeds. The production of machines in Austria ceased altogether in 1987 when the two-wheeler department of Steyr–Daimler–Puch was acquired by Piaggio.

1934 PUCH S4

The S4 was a popular sports bike built by Puch between 1934 and 1938. It used a two-stroke single-cylinder engine and was quite conventional for its day. Puch had some racing success with their two-stroke machines, including victory in the German Grand Prix of 1931 and although the road bikes were not of the same design, the two-stroke technology was efficiently applied.

SPECIFICATION
Country of origin: AUSTRIA
Capacity: 248cc (15.12cu. in.)
Engine cycle: 2-stroke
Number of cylinders: 1
Top speed: n/a
Power: 14.5bhp @ 4100rpm
Transmission: 4-speed
Frame: Tubular steel

1977 PUCH M50 CROSS

The M50 Cross was one of a range of three Puch machines with the designation M50, the others being the M50 Jet and M50 Grand Prix. The Jet and Grand Prix were roadgoing models while the Cross was an off-road, enduro-style motorcycle. All were based around 48.8cc (2.97cu. in.) engines and featured details specific to their style. Variations in types of brakes, mudguards and so on defined the styles.

SPECIFICATION
Country of origin: AUSTRIA
Capacity: 48.8 (2.97cu. in.)
Engine cycle: 2-stroke
Number of cylinders: 1
Top speed: 25mph (40kph)
Power: 2.6bhp @ 5500rpm
Transmission: 4-speed
Frame: Tubular steel

1980 PUCH MAXI S

The Maxi S was intended as basic transportation for the shopper market. It was a step-through design and it was available in three versions: the S which had a single-speed transmission; the D with a dual seat; and the N with a rigid frame.

SPECIFICATION
Country of origin: AUSTRIA
Capacity: 48.8cc (2.97cu. in.)
Engine cycle: 2-stroke
Number of cylinders: 1
Top speed: 28mph (45kph)
Power: 2.2bhp @ 4500rpm
Transmission: Single-speed
Frame: Step-through

ABOVE: The Pope company manufactured machines with V-twin engines, such as this model, which featured overhead valves rather than the inlet-over-exhaust configuration used by many of its contemporaries.

R

RADCO

Radco were a small company, based in England, that flourished between the two World Wars after being founded by E. A. Rad. They manufactured medium capacity machines such as a 211cc (12.87cu. in.) and 247cc (15.06cu. in.) two-strokes that featured external flywheels. Later the company went on to produce 247cc (15.06cu. in.) overhead-valve models with an engine of their own design. Like many other British manufacturers, Radco bought in two-stroke engines from Villiers. In this company's case it was 145 and 198cc (8.84 and 12.07cu. in.) units. The largest capacity machines from Radco were the 490cc (29.89cu. in.) JAP-engined singles which were produced in three different forms: Touring, Sport and Supersport. Production ceased in 1932 when the company began manufacturing only components. In 1958 they returned to two-wheeler production and built the RM1, a machine which featured a 49.5cc (3.01cu. in.) Sturmey Archer two-stroke engine. This motorcycle was offered in slightly modified form for a further two years until Radco switched to selling imported machines in the 1960s.

1922 RADCO 250

Radco produced a range of two-strokes including both 211 and 247cc (12.87 and 15.06cu. in.) models with external flywheels, and later went on to build machines of the latter displacement in overhead-valve form with an engine of the company's own

ABOVE: The Radco 250 of 1922 was a two-stroke with an external flywheel. The engine was fitted to a bicycle-type diamond frame and as such did not feature rear suspension. Later Radco 250s had overhead valves.

design. Both primary and final drive were by means of chains. The machine did not incorporate rear suspension and its frame was still manufactured along bicycle lines.

SPECIFICATION
Country of origin: GREAT BRITAIN
Capacity: 247cc (15.06cu. in.)
Engine cycle: 2-stroke
Number of cylinders: 1
Top speed: n/a
Power: n/a
Transmission: 2-speed
Frame: Steel diamond

RALEIGH

This English company, founded in 1899, are more noted for bicycle manufacture and the marque has now been absorbed into a large conglomerate. Motorcycles were built from 1902 until 1906. Later, after 1919, motorcycles were once again made in the Nottingham, England, factory.

They manufactured engines and gearboxes which they sold to other companies under the Sturmey Archer name. The 1930 range of Raleigh motorcycles included a number of vertical side-valves and some overhead-valves. The 1931 range was similar with models listed as the MO, MA, MT and MH which had engines with cylinders inclined forwards, a rear-mounted magneto, and redesigned frames and forks. Engine displacements ranged from 225 to 598cc (13.72 to 36.47cu. in.). In 1933 the company stopped two-wheeler production, preferring instead to concentrate on the manufacture of three-wheeler cars and vans. Production of these had ceased by 1935 and the company returned solely to bicycle manufacture.

1931 RALEIGH MH-31

Raleigh redesigned much about their motorcycles for the 1931 season, although many of the previous year's model designations were retained, including the MH. For 1931 the MH was fitted with an inclined overhead-valve engine with a rear-mounted magneto and a frame redesigned

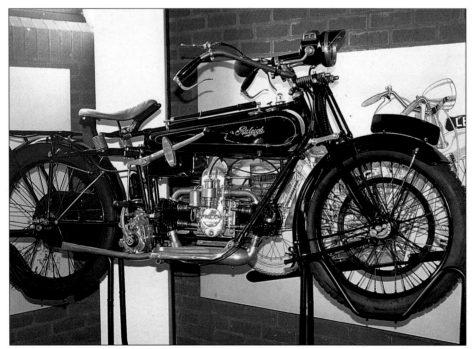

LEFT: Raleigh's heyday was during the 1920s when they produced a range of singles and flat-twins in varying displacements. They also built racing bikes with overhead-valve engines and a 798cc (48.67cu. in.) side-valve V-twin, such as the model shown.

to accept this. The forks were also modernized and the tank was chrome-plated.

SPECIFICATION
Country of origin: GREAT BRITAIN
Capacity: 495cc (30.19cu. in.)
Engine cycle: 4-stroke
Number of cylinders: 1
Top speed: 75mph (120kph)
Power: n/a
Transmission: 3-speed
Frame: Tubular steel

READING STANDARD

This American make flourished in Reading, Pennsylvania, during the first decades of the 20th century. Their first machines were Thor-engined machines and were not dissimilar to Indians of the time. The motorcycles were typical of American machines of the period. Production of both singles and V-twins was carried out, with singles of 499cc (30.43cu. in.) and V-twins of 990 and 1170cc (60.39 and 71.37cu. in.). The factory also produced race bikes for professional riders.

In 1907 Reading Standard were the first US maker to build side-valve engines. These were designed by Charles Gustafson who later moved to Indian where he designed the famous Powerplus engine. Reading Standard went out of business in 1922 and was purchased by Cleveland.

REGENT UK

The Regent range of commuter motorcycles are assembled in the UK by the importer after manufacture in the Republic of Belarus. The motorcycles are based around a 125cc (7.62cu. in.) two-stroke engine although both 150 and 175cc (9.15 and 10.67cu. in.) models are also manufactured. A range of styles including roadster, trail and a tourer are produced.

1995 REGENT ROADSTER

This motorcycle is intended as cheap, basic commuter transport based around an eastern European engine and components that are assembled in the UK. The motorcycles feature 12 volt electrics and electronic ignition. They are cheap to buy, cheap to run and uncomplicated to maintain.

BELOW: **The Regent Roadster is a motorcycle assembled (rather than manufactured) in the UK from parts made in Eastern Europe. The components are imported into Britain and re-finished prior to assembly and sale.**

SPECIFICATION
Country of origin: GREAT BRITAIN
Capacity: 125cc (7.62cu. in.)
Engine cycle: 2-stroke
Number of cylinders: 1
Top speed: n/a
Power: n/a
Transmission: 4-speed
Frame: Tubular cradle

RENÉ–GILLET

This French company was founded in 1898 and manufactured motorcycles until 1957. In this time many of their machines were used by the French military and also by the police. René–Gillet manufactured 748 and 996cc (45.62 and 60.75cu. in.) V-twins which were eminently suitable for pulling sidecars. In 1928 the company produced a 346cc (21.10cu. in.) side-valve single with sprung forks and a rigid frame.

In the years following World War II the company manufactured updated versions of its V-twins but shifted the majority of its production to two-stroke manufacture. The range of two-stroke models included 48, 123 and 246cc (2.93, 7.50 and 15cu. in.) machines. For all of these motorcycles the engines were of the company's own design and manufacture.

ABOVE: Rex Acme was the name given to the resulting entity after the amalgamation of Rex and Acme. The new company made machines such as this that still used proprietary engines.

REX

This English company was founded in 1900 and flourished until 1933. It produced both singles and V-twins and, after ceasing the manufacture of their own motorcycle engines, Rex began to use Blackburne units. Their early models featured a variety of quite innovative manufacturing techniques including a 372cc (22.69cu. in.) machine that had the silencer cast in one piece with the cylinder barrel.

The first Rex manufactured after World War I was a 550cc (33.55cu. in.) side-valve single but this was dropped in favor of motorcycles with Blackburne engines. Rex supplied their machines in two wheelbases

ABOVE: In the years after World War I Rex moved from using engines of their own design to those bought in from companies such as Blackburne, a supplier of proprietary engines to numerous motorcycle manufacturers.

depending on whether or not a sidecar was to be attached. In 1922 Rex amalgamated with Acme to form Rex Acme. Both companies were based in Coventry, England. The new company's products soon achieved riding success with Walter Handley on board. He ultimately became a director and then later left in order to race on other marques. The Depression was not kind to the company and it only survived by manufacturing machines from bought-in parts. The factory

was taken over by a sidecar manufacturer in 1932 who stopped motorcycle production the following year.

ROVER

This English manufacturer is perhaps better known for the production of cars but it also built motorcycles between 1902 and 1925. They were reputed to be well made and finished. Prior to 1914 the company manufactured a 496cc (30.25cu. in.) side-valve single and it continued to be built in the years following World War I. In common with some other motorcycle makers, Rover also used proprietary engines such as a JAP V-twin of 676cc (41.23cu. in.). During the 1920s Rover produced 248 and 348cc (15.12 and 21.22cu. in.) singles of their own design until all motorcycle production was halted in favor of car manufacture.

BELOW: The last Rover motorcycles built used single-cylinder engines of the company's own design and manufacture. They were available in 248 and 348cc (15.12 and 21.22cu. in.) between 1923 and 1925. This is the larger displacement machine of 1925.

ROYAL ENFIELD

In the closing years of the 19th century the Enfield Cycle Company began work on three- and four-wheelers with proprietary engines fitted from companies such as De Dion, Minerva and MMC. The company manufactured their first motorcycle in 1900 and continued until 1905 when motorcycle production stopped. It resumed five years later when an MAG-engined V-twin was made. This was soon followed by a 770cc (46.97cu. in.) Jap-engined V-twin. Enfield then went on to produce motorcycles with their own designs of engine.

In the years after World War I the company concentrated on production of a big V-twin of 976cc (59.53cu. in.) displacement. A medium weight four-stroke was introduced in 1924, equipped with a JAP engine although the company later installed an engine of their own design and manufacture. A 996cc (60.75cu. in.) V-twin outfit ridden by a Swede, E. Magner, broke the world record for sidecars over a distance of one mile (1.6km).

In 1928 the Redditch, England, company adopted saddle tanks and center-spring girders and were one of the first to adopt swinging arm rear suspension after World War II. From 1930 onward Royal Enfield machines were given an alphabetical designation. The Model A, for example, was a 225cc (13.72cu. in.) two-stroke. The Model Z Cycar was an unusual 148cc (9.02cu. in.) machine introduced in 1934. The range was adjusted throughout the 1930s and vertical singles appeared in 1936. By 1938 there were 20 different models in the Royal Enfield range. World War II intervened, during which Royal Enfield manufactured the Flying Flea. The company also built a 346cc

ABOVE: **This 1925 Royal Enfield was made at the time when the company were shifting the emphasis of their production away from being reliant on proprietary engines.**

(21.10cu. in.) Model C and CO singles for the Army. In the postwar years the company produced a range of twins and singles. Of the latter the Bullet, in two capacities of 350 and 500cc (21.35 and 30.50cu. in.), is famous partially because it is still manufactured in India as the Enfield India. The Twins included models such as the Super Meteor, Crusader and Continental. Norton–Villiers acquired the company when it collapsed in 1967.

1941 ROYAL ENFIELD FLYING FLEA

The Flying Flea came about at the request of a Dutch company who had sold German DKW motorcycles in Holland prior to World War II. The Jewish-owned company had its franchise removed by the Nazis and asked Royal Enfield to supply a substitute. The result was a lightweight 125cc (7.62cu. in.) lightweight. When the war finally arrived it was redesigned to accompany paratroopers into battle by parachute. It featured blade girder forks.

SPECIFICATION
Country of origin: GREAT BRITAIN
Capacity: 125cc (7.62cu. in.)
Engine cycle: 2-stroke
Number of cylinders: 1
Top speed: n/a
Power: n/a
Transmission: 3-speed
Frame: Tubular loop

BELOW: **Around 55,000 of the wartime Flying Flea were built by Royal Enfield for the British Airborne forces who took them by air into battle to give mobility once on the ground. Postwar they remained in production.**

ABOVE: The Royal Enfield model name, Bullet, was introduced in 1933 but the 1950s roadgoing 350cc model seen here featured telescopic forks.

1955 ROYAL ENFIELD BULLET

The Royal Enfield Bullet, a medium capacity, overhead-valve single was launched in 1949 in road, trials and scrambles form. Shown above is the 1955 350cc class model in road trim. It features swinging arm rear suspension. The machine was a basic commuter machine and later joined by a 499cc (30.43cu. in.) version suited for solo or side-car use. A redesign for 1954, which lasted several years, was the introduction of the headlamp nacelle. The Bullet is still manufactured in India as the Enfield India and surprisingly little altered from the 1950s British version.

SPECIFICATION
Country of origin: GREAT BRITAIN
Capacity: 346cc (21.10cu. in.)
Engine cycle: 4-stroke
Number of cylinders: 1
Top speed: 68mph (109kph)
Power: 18bhp
Transmission: 4-speed
Frame: Tubular cradle

1964 ROYAL ENFIELD CONTINENTAL GT

This 250 became popular as a café racer because of its sleek lines and also as a result of a number of its features which had been inspired by the race-track. These included the clip-on handlebars, drum brake cooling discs and the racing-style seat. Production of the Continental only stopped because of the eventual closure of Royal Enfield.

SPECIFICATION
Country of origin: GREAT BRITAIN
Capacity: 248cc (15.12cu. in.)
Engine cycle: 4-stroke
Number of cylinders: 1
Top speed: 86mph (142kph)
Power: 26bhp @ 7500rpm
Transmission: 4-speed
Frame: Tubular cradle

RUDGE

The established Rudge–Whitworth bicycle factory in Coventry, England, started producing motorcycles in 1911. Their first production machine was a 499cc (30.43cu. in.) inlet-over-exhaust configuration machine. In the years after the World War I the company produced a 749cc (45.68 cu. in.) single and a 998cc (60.87 cu. in.) V-twin. Rudge also built what became known as the multi-gear which was a variable gear that gave up to 21 positions. In 1923 and 1924 major changes were made to Rudge's range in terms of new engines and new gearboxes. By the 1930s their reputation was firmly established as makers of fine motorcycles, no doubt helped by some racing success. An example of this was a win in the 1928 Ulster Grand Prix which led to a 499cc (30.43cu. in.) sports model being called the Rudge Ulster. In 1930 at the Isle of Man TT Rudge won both Senior and Junior events. The Depression adversely affected Rudge's sales of complete motor-

BELOW: Rudge listed the Ulster in their 1929 catalog. In 1930 one of these machines was ridden to first place in the Isle of Man Senior TT by Walter Handley.

RIGHT: Rudge built a model designated the Special through the whole of the 1930s. It was an overhead-valve design with four valves.

cycles and they offered their engines and gearboxes to other manufacturers under the Python name. Racing remained a priority at Rudge despite indifferent sales. Eventually in 1935 they were acquired by EMI and the works moved to Hayes, England, from Coventry. EMI helped Rudge invest in the autocycle market but during World War II the company was sold to Raleigh industries while the factory worked on radar projects.

1912 RUDGE SINGLE

The long established bicycle factory entered motorcycle production in 1911. The company's first model was an inlet-over-exhaust single based on a bore and stroke of 85 and 132mm. The fuel tank fitted in between the rails of the bicycle-type frame. Final drive was by means of a belt.

SPECIFICATION
Country of origin: GREAT BRITAIN
Capacity: 499cc (30.43cu. in.)
Engine cycle: 4-stroke
Number of cylinders: 1
Top speed: n/a
Power: n/a
Transmission: Single-speed
Frame: Tubular steel

1938 RUDGE ULSTER

Success at the Ulster Grand Prix of 1928 led Rudge to name one of their models after the Northern Irish venue. It was a name which endured from that model year until the outbreak of World War II. The Ulster was gradually upgraded year by year, and it survived the company's liquidation in 1933 and subsequent acquisition by EMI. The later Ulsters, such as the 1938 model, featured a semi-radial four-valve engine – the valve gear was enclosed only the year before. The chassis of the machine included a rigid frame and girder forks in which were mounted drum-braked spoked wheels.

SPECIFICATION
Country of origin: GREAT BRITAIN
Capacity: 499cc (30.43cu. in.)
Engine cycle: 4-stroke
Number of cylinders: 1
Top speed: n/a
Power: n/a
Transmission: 4-speed
Frame: Tubular steel

BELOW: By 1938 the Rudge Ulster, as shown in this photograph, featured enclosed valve gear and semi-radial four valves in its cylinder head. The rockers were covered with an alloy cover. The outbreak of World War II put an end to its production.

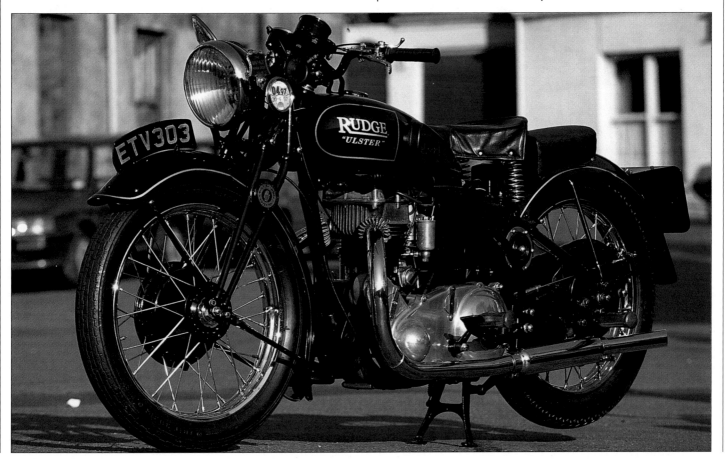

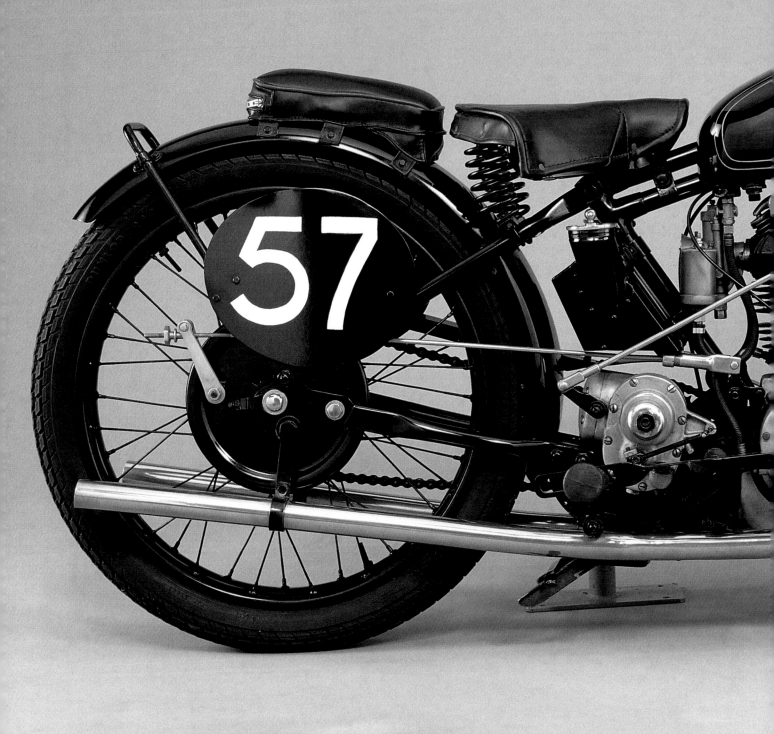

ABOVE: Rudge started production of their TT replica in 1937 by which time the company had been racing for many years. The overhead-valve machine produced 32bhp and was capable of speeds of up to 100mph (161kph).

S

SANGLAS

This Spanish company was founded in 1942 and built a range of four-stroke, overhead-valve, single-cylinder machines with unit construction engines and gearboxes. These were made in capacities which ranged from 295 to 497cc (17.99 to 30.31cu. in.). These were constructed for the Spanish police forces as well as other customers. In the 1950s the company continued with four-stroke production which made them somewhat unusual as most Spanish companies produced lightweight two-strokes. Sanglas did build some two-strokes in the early 1960s using Zündapp and Villiers engines badged as Rovenas. Sanglas exported its machines to some South American countries. Production of two-strokes was dropped in 1968 and by 1973 the company had launched the 400E which was later upgraded into the 400F, a 423cc (25.80cu. in.) single. Spain had a policy of not allowing the importation of Japanese bikes and Sanglas built the 400Y. This was Spanish-designed and built but had a Yamaha XS engine. Yamaha acquired Sanglas in 1981 and dropped the Sanglas name.

1981 SANGLAS 500S ZV5

This was the largest capacity version of Sanglas' big single-cylinder, four-strokes that was launched in 1973. The models were gradually upgraded and exported, particularly where there was Spanish influence such as certain South American countries. It was a motorcycle typical of its time with spoked alloy wheels, disc brakes and its sporting aspirations were indicated by a bikini fairing and a dual seat that led into the tailpiece.

SPECIFICATION
Country of origin: SPAIN
Capacity: 496cc ((30.25cu. in.)
Engine cycle: 4-stroke
Number of cylinders: 1
Top speed: 99mph (160kph)
Power: 35bhp @ 6700rpm
Transmission: 5-speed
Frame: Duplex tubular cradle

SAROLEA

Joseph Sarolea, who was an arms manufacturer, founded this Belgian marque in 1901. His first machines were powered by proprietary engines of both single and V-twin types. In the years between the two World Wars the company produced engines in both side- and overhead-valve configurations. It went on to produce a range of machines which included the sporting Monotube, a functional two-stroke, and a flat-twin sidecar outfit of 980cc (59.78cu. in.) with a driven sidecar wheel, for the Belgian military. After World War II certain prewar models were reintroduced and in 1950 a 125cc (7.62cu. in.) two-stroke was introduced, although a 498cc (30.37cu. in.) parallel twin was also produced. In 1960 Sarolea merged with French Gillet and production stopped in 1963.

SCHÜTTOFF

This German machine factory in Chemnitz started motorcycle production in 1924 and manufactured a range of conventional machines of medium capacity. They started with a 246cc (15cu. in.) side-valve which was later enlarged to 346cc (21.10cu. in.) and built in side- and overhead-valve configurations. A 496cc (30.25cu. in.) machine was also made. From 1930 they used proprietary two-stroke engines from the DKW company. These were of both 198 and 298cc (12.07 and 18.17cu. in.) displacement. In 1932 DKW acquired the Schüttoff concern. During the period of their manufacture, Schüttoff machines had some racing success, most notably with the 348cc (21.22cu. in.) overhead-valve Model Sport in the late 1920s.

1930 SCHÜTTOFF/DKW

This machine reflects the close co-operation between the Schüttoff and DKW concerns. DKW had purchased a stake in the company in 1928. It uses a four-stroke, overhead-valve, single-cylinder engine and was badged as a DKW and assembled in the DKW works at Zschopau.

SPECIFICATION
Country of origin: GERMANY
Capacity: 500cc (30.50cu. in.)
Engine cycle: 4-stroke
Number of cylinders: 1
Top speed: 68mph (109kph)
Power: 14PS
Transmission: 3-speed
Frame: Tubular steel

SCOTT

This company was founded by Alfred A. Scott in 1909 in Shipley, England, and was

BELOW: A. A. Scott was a pioneer of both the two-stroke motorcycle and the concept of water-cooling due to his ability to combine advanced ideas and sound engineering.

1933 SCOTT FLYING SQUIRREL

The water-cooled two-stroke engine used in the Flying Squirrel was unusual in its design; the flywheel was centrally positioned with twin inboard main bearings, overhung crankpins and crankcase doors to allow access to the engine. The whole was located in a large alloy casting on which sat the block with non-detachable cylinder heads. The engine was made in two capacities of 499 and 597cc (30.43 and 36.41cu. in.) through a common stroke but change in bore diameter. The fuel tank was positioned atop the triangulated frame and was oval in shape. A number of variants including sprint and TT replica models of the Flying Squirrel were offered. The engines drove two- and three-speed gearboxes. The TT replica models used specially constructed long stroke Power Plus engines based on the TT winning machines of 1928.

SPECIFICATION
Country of origin: GREAT BRITAIN
Capacity: 499cc (30.43cu. in.)
Engine cycle: 2-stroke
Number of cylinders: Twin
Top speed: n/a
Power: n/a
Transmission: 3-speed
Frame: Triangulated duplex

among the pioneers of the two-stroke motorcycle. They had some Isle of Man TT success including wins in 1912 and 1913. Most Scott motorcycles were water-cooled and Scott is reputed to have had a flair for combining unorthodox ideas with sound and practical engineering. Scott had rotary inlet valves, water-cooling and the 180° parallel-twin two-stroke engine in the early years of the 20th century yet many of these ideas have only been adopted by Japanese manufacturers much more recently. Scott utilized the engine as a stressed member in his duplex, triangulated tube frames which enhanced handling. He also a pioneered all chain drive and countershaft gears. A. A. Scott relinquished his connection with the company in 1915 and died in 1923. Production of his basic designs continued.

ABOVE: An early 1930s 299cc (18.23cu. in.) single-cylinder Scott. The two-stroke engine was of Scott's own design rather than the proprietary engine from Villiers.

BELOW: A 1933 Scott Flying Squirrel. By this time the two-stroke Scott was considered an anachronism but the company failed to develop more advanced machines.

1931 SCOTT SINGLE

From 1929 Scott built an air-cooled single as a bike for the smaller end of the market. Unlike most other British manufacturers they did not purchase a Villiers proprietary engine but designed their own. It was a crude design with an iron barrel and alloy head. The number of head bolts had to be increased in 1930 from three to six as the smaller number allowed the head to warp. The single was equipped with girder forks, a duplex frame and magneto ignition.

SPECIFICATION
Country of origin: GREAT BRITAIN
Capacity: 299cc (18.23cu. in.)
Engine cycle: 2-stroke
Number of cylinders: 1
Top speed: n/a
Power: n/a
Transmission: 3-speed
Frame: Duplex tubular

1950 SCOTT FLYING SQUIRREL

Scott did not manufacture motorcycles during World War II but started producing them immediately after its end. The first model to be reinstated was the Flying Squirrel. Although it featured new brakes, the machine was at first produced with girder forks, although these were soon replaced by telescopic forks. The ignition was upgraded in 1949 when coil ignition was fitted and a roll-on centerstand was now utilized. Many other aspects of the Flying Squirrel were considered dated and it was an expensive machine. These two factors meant that sales were poor and the company went into voluntary liquidation. The result of this was that the company had to move from its Yorkshire premises to Birmingham under the auspices of new owners. The 1950 Squirrel was one of the last Scotts made in the original works.

SPECIFICATION
Country of origin: GREAT BRITAIN
Capacity: 596cc (36.35cu. in.)
Engine cycle: 2-stroke
Number of cylinders: 2
Top speed: 85mph (140kph)
Power: 30bhp @ 5000rpm
Transmission: 3-speed
Frame: Tubular steel

ABOVE: The postwar Flying Squirrel was soon updated through the fitting of Dowty telescopic forks, shown on this 1950 model. Also new for 1950 was the roll-on centerstand, separate oil tank as a result of a shift from magneto to dynamo, and coil ignition.

SEARS

Sears, Roebuck & Company, the department store which was based in Chicago, Illinois, sold motorcycles branded as Sears that were made for it by other concerns several times throughout its history. The first time it did this was in 1912 when single-cylinder motorcycles were listed in the company's catalog. The simple machines featured a Spacke single-cylinder inlet-over-exhaust engine with a Bosch magneto and chain transmission. The whole motorcycle was based around a loop frame and leaf sprung trailing link forks. Sears also offered a machine with a V-twin engine. Both of these models were dropped in 1916. Later Sears sold Austrian Puch machines and Italian Gileras, all branded as Sears' products. For a while the company concentrated on lightweights and scooters (some of which were sold as Allstates) but Sears dropped out of motorcycle retail completely in 1968 when Japanese competition began to get stronger and better organized.

SILK

This company, which was founded in 1974 in Great Britain, grew out of a Scott restoration business. It was established by George Silk and Maurice Patey in order to build a Scott-inspired but redesigned and freshly-engineered motorcycle. The new machine was designed by George Silk and David Midgelow.

1977 SILK 700S MK II SABRE

A limited production model, the Silk 700S was a sports bike with an all-aluminum, water-cooled, two-stroke engine. The compact and lightweight engine together with a compact lightweight frame gave a motorcycle that was renowned for good handling. It was also constructed to a high specification through the use of stainless steel components and transistorized ignition.

SPECIFICATION
Country of origin: GREAT BRITAIN
Capacity: 653cc (39.83cu. in.)
Engine cycle: 2-stroke
Number of cylinders: 2
Top speed: 111.8mph (180kph)
Power: 48bhp @ 6500rpm
Transmission: 4-speed
Frame: Tubular cradle

SIMPLEX

This American company produced simple lightweight machines from 1935. The Servicycle as it was known was successful and enjoyed a long production run with only minor upgrades during that time. (An automatic clutch was introduced in 1953, for example.) The firm also made scooter models while these were fashionable, and later built a lightweight two-wheeler powered by a two-stroke Clinton industrial engine. Other manufacturers in Italy, Holland and England used Simplex as a brandname. Of these, the Dutch concern was the largest and produced a wide range of motorcycles and mopeds under the Simplex name in the years between 1902 and 1968.

SIMSON

This East German company was a former arms factory which started building shaft-drive 246cc (15cu. in.) AWO machines from 1949, although the Simson brothers had made bicycles as early as 1898 in the German IFA group of companies. The AWO machines were made until the 1950s when the factory concentrated on the mass production of 74 and 49cc (4.51 and 2.98cu. in.) mopeds. In the late 1960s the company built a range of off-road machines including those for the East German ISDT team.

1980 SIMSON S50 B1

The range that comprised the S50 was based on a design that was introduced in 1975. It was a fully conventional two-stroke moped intended simply as basic transporta-

ABOVE: The Simson SR50/1C scooter of 1991 was a new style of machine from the company. The SR50/1C was both step-through and scooter, based around a single-cylinder, two-stroke engine and small diameter wheels.

tion. Since the 1950s over 3 million Simson lightweights have been made in the factory in Suhl, Germany.

SPECIFICATION
Country of origin: GERMANY
Capacity: 49.6cc (3.02cu. in.)
Engine cycle: 2-stroke
Number of cylinders: 1
Top speed: 37mph (60kph)
Power: 3.6bhp @ 5500rpm
Transmission: 3-speed
Frame: Tubular steel

SUNBEAM

Sunbeam was founded in 1912 by John Marston who originally built bicycles in his Wolverhampton, England, works using the Sunbeam trademark. The first Sunbeam motorcycle was constructed with a 347cc (21.20cu. in.) side-valve engine, two-speed gearbox and enclosed primary and final drive chains. Later on the company used engines from AKD, JAP and MAG and supplied motorcycles with V-twin engines. The company gained a reputation for excellence and was acquired by ICI, a chemical company, in 1918. The Isle of Man TT saw four Sunbeam wins in the races of 1920, 1922, 1928 and 1929. During this time the factory produced ohv singles but by the late 1920s the roadgoing machines were not faring well against other British makes and the pressure was on to compete more effectively. A new range of 250 and 500cc (15.25 and 30.50cu. in.) models were introduced in the early 1930s. Times were hard and Sunbeam's reputation for quality was adversely affected by a cost-cutting operation. Profits were still not sufficient for owners ICI so the company was sold to the AJS and Matchless concern in London. After this acquisition the company became AMC and production moved to London. World War II intervened and in 1943 AMC sold the Sunbeam name to BSA. This resulted in a totally new line up of postwar twins, the 487cc (29.70cu. in.) shaft drive machines. In the mergers and contraction of the last years of the British motorcycle industry, the Sunbeam name appeared on BSA scooters but disappeared completely in 1964.

BELOW: The Sunbeam Model 8 from 1913 with sidecar. The machine was fitted with a JAP 346cc (21.10cu. in.) side-valve engine. This was one of the earliest machines from the company, founded the previous year.

ABOVE: A side-valve Sunbeam from the late 1920s. Soon after this the company switched to saddle tanks, although it was already offering both side- and overhead-valve models.

1928 SUNBEAM MODEL 90

Aboard one of the Sunbeam machines Charlie Dodson won the senior Isle of Man TT in both 1928 and 1929. Sunbeam introduced overhead-valve machines in 1923 and of these the 500cc model was the most successful. While the Model 90 was primarily a racing bike, it was also offered in road-going trim with additional features such as a kick starter, enclosed rear chain and lights. The flat tanks of the first models were dropped for 1929 when the more modern saddle tank was introduced.

SPECIFICATION
Country of origin: GREAT BRITAIN
Capacity: 493cc (30.07cu. in.)
Engine cycle: 4-stroke
Number of cylinders: 1
Top speed: 90mph (150kph)
Power: n/a
Transmission: 3-speed
Frame: Steel tubular

1935 SUNBEAM LION

The Lion was introduced into the Sunbeam range in June 1930 two years after ICI had acquired the company. The new machine featured Webb forks and a chrome-plated fuel tank and it was powered by a 489cc (29.82cu. in.) side-valve engine. Changes were made to it for the following years in order to trim costs as the company faced the Depression. A 599cc (36.53cu. in.) version was introduced in 1932. By 1935 the Lion was still being made in the two displacements.

BELOW: A 1935 Sunbeam lion of 599cc (36.53cu. in.), the larger of two displacements built in that year. Changes to the transmission were made for 1936.

SPECIFICATION
Country of origin: GREAT BRITAIN
Capacity: 599cc (36.53cu. in.)
Engine cycle: 4-stroke
Number of cylinders: 1
Top speed: n/a
Power: n/a
Transmission: 4-speed
Frame: Tubular

1951 SUNBEAM S7

The Sunbeam S7, and the variant S8, were quite rare examples of the truly new post-

war British motorcycles and were designed by Erling Poppe. They were introduced in 1947 with an in-line, twin-cylinder engine, four-speed gearbox, shaft drive to the rear wheel and telescopic forks. Some problems with the new design caused improvements to be made during the production run. The S8 was a lighter version and production of both models lasted until 1958.

SPECIFICATION
Country of origin: GREAT BRITAIN
Capacity: 489cc (29.82cu. in.)
Engine cycle: 4-stroke
Number of cylinders: 2
Top speed: 75mph (120kph)
Power: 25bhp
Transmission: 4-speed
Frame: Duplex plunger

SUZUKI

Suzuki was originally a textile engineering company that had been founded in 1909 by Michio Suzuki. The company started its production of motorcycles in 1952 when it built the 36cc (2.19cu. in.) Power Free motorized bicycle. Based in Hamamatsu, Japan, the company changed its name to Suzuki Motor Company Limited in 1954 and then entered a period of phenomenal growth. As well as two-stroke engines the company produced outboard motors, light 4x4s, cars, bicycles, motorboats and even prefabricated housing units.

ABOVE: Suzuki began racing in the 1960s in the smaller capacity classes on machinery developed for that purpose; one such model was the 1967 RS67, a 125cc (7.62cu. in.) works racer powered by a V4 engine.

The company started racing in Europe in 1960 in the smaller classes but by 1970 it was racing full works' teams with riders such as Barry Sheene aboard the RG500. In the late 1970s Suzuki won the manufacturers' world 500cc championship four times in a row. Suzuki initially concentrated on two-stroke production but was forced to add four-strokes to its range due to the increasingly stringent emissions regulations in the USA, one of Suzuki's largest export markets. Suzuki, in common with the other Japanese manufacturers, has always considered exports an important part of its business, particularly to Europe and the USA. It has plants in Malaysia, Taiwan, Indonesia, the Philippines, Mozambique and Nigeria.

BELOW: The 1973 Suzuki TR 750 on which Barry Sheene won the FIM Formula 750 title of that year. It had a three-cylinder engine fitted into a Seeley frame to enhance handling.

1975 SUZUKI GT750M

The GT750 was a liquid-cooled, two-stroke triple with a 6.9:1 compression ratio and three Mikuni carburetors. It was typical of its time with disc front brakes and a drum rear spoked wheel, a dual seat and a large

BELOW: A 1976 Suzuki GT750, this was a water-cooled two-stroke of 738cc (45.01cu. in.). Also in that year's range was the GS750, a four-stroke, four-cylinder motorcycle.

fuel tank. The GT750 was gradually upgraded as the 1977 model shows.

SPECIFICATION
Country of origin: JAPAN
Capacity: 738cc (45.01cu. in.)
Engine cycle: 2-stroke
Number of cylinders: 3
Top speed: 119mph (192kph)
Power: 70bhp @ 6500rpm
Transmission: 5-speed
Frame: Tubular cradle

1976 SUZUKI RG500

This two-stroke race bike took Barry Sheene to World Championships in 1976 and 1977. It also saw numerous Isle of Man TT wins – John Williams in 1976, as well as other victories from Mick Grant, Mike Hailwood and Phil Read. The engine was a square four which fired in diagonal pairs every 180°. Suzuki had gained considerable knowledge of two-strokes when Erst Degner, MZ's leading rider, had defected from East Germany taking many of Walter Kaaden's secrets with him.

SPECIFICATION
Country of origin: JAPAN
Capacity: 497cc (30.31cu. in.)
Engine cycle: 2-stroke
Number of cylinders: 4
Top speed: n/a
Power: 100bhp @ 10,000rpm
Transmission: 6-speed
Frame: Tubular cradle

BELOW: In the early 1980s the British Suzuki importer Heron–Suzuki developed an aluminum-framed RG500 for racing. It was ridden by Rob McElnea, whose RG500 racer is shown here. The race-bred RG500 spawned a road bike, the RG500 Gamma, a fast 95bhp, two-stroke machine.

1977 SUZUKI GT750M

The GT750 was launched in 1971 as a water-cooled, two-stroke, in-line triple-cylinder motorcycle. It was relatively sophisticated for its time and updated throughout its production run. The twin disc front brakes seen on the 1977 model, for example, were instituted in 1974 to replace the four leading shoe drum originally fitted. Due to its water-cooling, the GT750 was nicknamed the 'Kettle' in Britain and the 'Water Buffalo' in the USA . Emission regulations outlawed it in the USA after 1977.

SPECIFICATION
Country of origin: JAPAN
Capacity: 738cc (45.01cu. in.)
Engine cycle: 2-stroke
Number of cylinders: 3
Top speed: 119mph (192kph)
Power: 70bhp @ 6500rpm
Transmission: 5-speed
Frame: Tubular cradle

ABOVE: The two-stroke engine in the GT750M did not meet tough US emissions legislation in the late 1970s leading Suzuki to follow other Japanese makers in introducing four-strokes.

1977 SUZUKI RE5

1977 was the last year in production for the RE5, which had only been introduced in 1974. It featured a Wankel Rotary engine and was innovative for its time. Unusual features included a cylindrical instrument binnacle which rotated open when the ignition was switched on. In many other ways it was conventional, being based on a tubular steel swinging arm frame and telescopic forks. A later variant, the RE5A, used a number of GT750 cycle parts.

SPECIFICATION
Country of origin: JAPAN
Capacity: 497cc (30.31cu. in.)
Engine cycle: Rotary
Number of cylinders: Single chamber
Top speed: 109mph (176kph)
Power: 62bhp @ 6500rpm
Transmission: 5-speed
Frame: Tubular cradle

LEFT: The RE5 Rotary-engined Suzuki had a short-lived production run from 1974, when this model was made, to its discontinuation in 1977 after a redesign into the RE5A.

1977 SUZUKI GS750

The four-stroke range introduced by Suzuki to ensure success in its export markets were given the GS designation while the numerical part of it approximated to the metric displacement of the engine. Suzuki produced GS1000, 850, 750, 500 and 400 models over a number of years. They had similar conventional styling and were upgraded during production. One change was from wire spoked to cast alloy wheels, for example. The GS750 was the fastest 750 on the market when introduced in 1976.

BELOW: The 1980 GS1000, four-cylinder, four-stroke motorcycle, was marketed in two forms: the GS1000E, shown here; and the GS1000S, which was almost identical but featured bikini sports fairing and different paint schemes.

SPECIFICATION
Country of origin: JAPAN
Capacity: 748.7cc (45.67cu. in.)
Engine cycle: 4-stroke
Number of cylinders: 4
Top speed: 124mph (200kph)
Power: 68bhp @ 8500rpm
Transmission: 5-speed
Frame: Duplex tubular cradle

1980 SUZUKI GSX1100

Another designation from Suzuki was GSX. This identified a double overhead camshaft, sixteen-valve, four-stroke engine and was produced in various capacities. The GSX 1100 was the largest of the models when it was introduced and featured cast alloy wheels, disc brakes back and front, and high performance.

SPECIFICATION
Country of origin: JAPAN
Capacity: 1075cc (65.57cu. in.)
Engine cycle: 4-stroke
Number of cylinders: 4
Top speed: 125mph (201kph)
Power: n/a
Transmission: 5-speed
Frame: Steel cradle

1986 SUZUKI GSX1100S KATANA

Like many successful models, the designation was retained for an altogether uprated motorcycle. In 1982 Suzuki produced the Katana range of machines which featured angular styling of both fuel tank and fairing. The GSX1100S was the 1986 variant, available in various displacements including 997 and 1074cc (60.81 and 65.51cu. in. respec-

tively). It included such design features as anti-dive forks, disc brakes and alloy wheels. It was intended as a high performance sports bike.

SPECIFICATION
Country of origin: JAPAN
Capacity: 1074cc (65.51cu. in.)
Engine cycle: 4-stroke
Number of cylinders: 4
Top speed: 144mph (232kph)
Power: 111bhp @ 8500rpm
Transmission: 5-speed
Frame: Duplex tubular cradle

1991 SUZUKI GSX-R1100

While the GSX designation was retained on a range of four-cylinder Suzukis, a new designation was also derived from it. This new model was the GSX-R and when launched in 1986 heralded the start of the true race-

BELOW: The GSX-R1100 of 1990 featured works-racer styling with technology to match. A full fairing assisted in engine-cooling through what Suzuki described as SACS (Suzuki Advanced Cooling System) which used oil. The bike is based around a cradle frame.

replica era from the big Japanese manufacturers. The GSX-R1100 was both fast and clearly race-inspired with its full fairing and race-type frame.

SPECIFICATION
Country of origin: JAPAN
Capacity: 1127cc (68.74cu. in.)
Engine cycle: 4-stroke
Number of cylinders: 4
Top speed: 167 mph (270kph)
Power: 143bhp @ 9500rpm
Transmission: 5-speed
Frame: Aluminum cradle

ABOVE: The GSX-R of 1985 was introduced alongside the RG500 Gamma as Suzuki shifted their emphasis to four-stroke machines. It was state-of-the-art for its day: liquid-cooled, four-valves per cylinder, 100bhp @ 10,500rpm.

ABOVE: By 1991 the GSX-R featured SACS cooling but it was still a four-valve per cylinder, four-cylinder, four-stroke which, in standard form, produced 120bhp @ 11,000rpm. It was a smaller version of the GSX-R1100.

ABOVE: The evolution of the GSX-R1100 continued and by 1996 the motorcycle was essentially the same but refined in detail. Now known as the GSX-R1100WT, it generated 126bhp @ 12,000rpm.

153

ABOVE: Rob McElnea (Number 19) was a British motorcycle racer who rode the four-cylinder, two-stroke RG500 Suzuki that was campaigned in various forms by the British Suzuki importers, Heron–Suzuki.

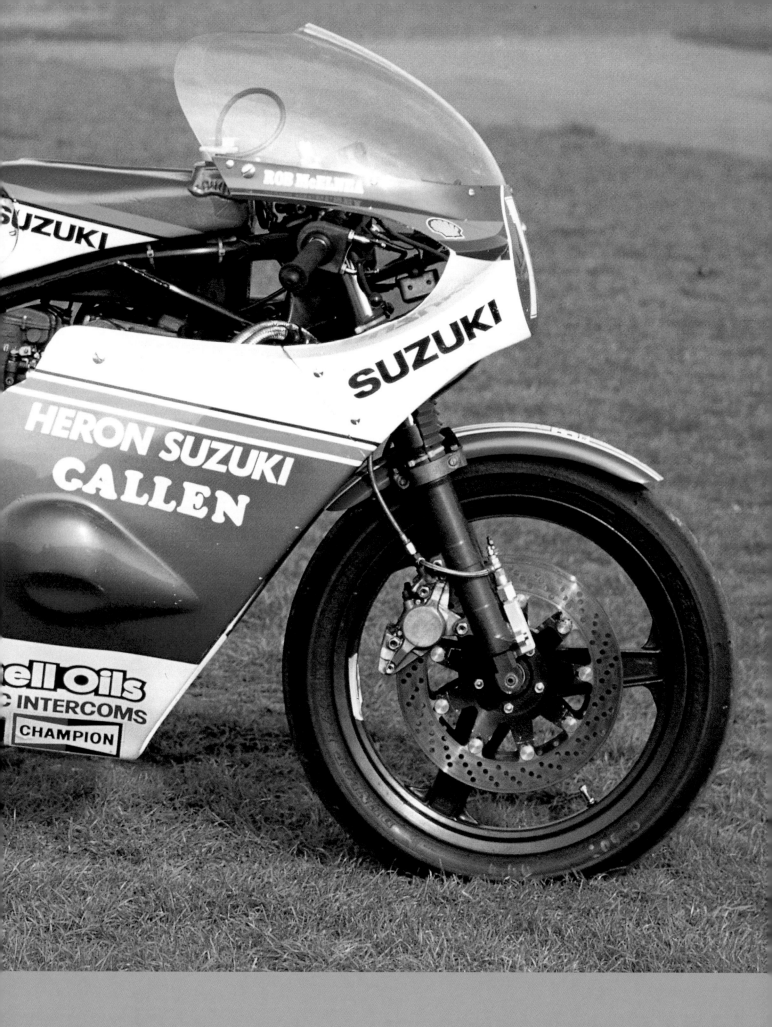

1991 SUZUKI VS750GLP INTRUDER

The Japanese factories started building US custom-styled motorcycles, a move which was no doubt inspired by the renaissance of the Harley–Davidson marque in the late 1980s. Suzuki built the Intruder range for this market. Like their other bikes it was built in a variety of displacements including the 750cc and the larger 1360cc variant, the VS1400GL. It featured a V-twin engine, shaft drive and an upright riding position as well as custom styling that included a teardrop fuel tank, pullback bars and a two-level seat.

SPECIFICATION
Country of origin: JAPAN
Capacity: 747cc (45.56cu. in.)
Engine cycle: 4-stroke
Number of cylinders: 2
Top speed: 105mph (170kph)
Power: 55bhp @ 7500rpm
Transmission: 5-speed
Frame: Steel cradle

BELOW: The VS750GLP Intruder of 1991 was a factory custom and featured a teardrop fuel tank, a stepped seat and pillion back rest, pullback handlebars and a V-twin engine with shaft drive from a five-speed gearbox.

1993 SUZUKI GSX-R1100W

The Suzuki GSX-R1100W was a large capacity sports bike which was first introduced in 1986 as the GSX-R1100G and has been sequentially upgraded each model year ever since. Over that period, a number of major improvements to the design have included a change of carburetor type for the GSX-R1100K, 'upside-down' forks for the J suffix models, changes to the headlamps and to the fairings for the M-suffix machines, and then a shift to water-cooling with the later models which included the WP, WR, WS and WT.

SPECIFICATION
Country of origin: JAPAN
Capacity: 1074cc (65.51cu. in.)
Engine cycle: 4-stroke
Number of cylinders: 4
Top speed: 177mph (284kph)
Power: 131bhp @ 9500rpm
Transmission: 5-speed
Frame: Aluminum cradle

1993 SUZUKI DR350

This machine is a dual purpose trail bike designed for both on- and off-road use. It is powered by a single-cylinder, four-valve engine. It features long travel suspension and disc brakes, front and rear. The high stance is enhanced by 18 and 21in. (450 and 525mm) wheels back and front respectively. The machine is assembled around a cradle frame that includes protection for the bottom of the engine for off-road riding.

SPECIFICATION
Country of origin: JAPAN
Capacity: 349cc (21.28cu. in.)
Engine cycle: 4-stroke
Number of cylinders: 1
Top speed: n/a
Power: 23.5bhp @ 7000rpm
Transmission: 6-speed
Frame: Tubular cradle

ABOVE: The 1991 GSX-R1100 was updated by turning it into a water-cooled version of the existing oil-cooled GSX-R1100 series. The upgrade involved a slight reduction in the displacement but an increase in top speed.

BELOW: The DR350 is a dual purpose machine designed for on- and off-road use. It features a high stance to assist in crossing uneven ground and is based around a single-cylinder engine that provides maximum torque at low revs.

BELOW: Race bikes such as Kevin Schwantz's 185mph (297kph) Suzuki RGV500 are specially built for racing. This 1992 machine has a twin crank, V-4, water-cooled, two-stroke engine with a bore and stroke of 56 and 50.6mm respectively, giving a capacity of 498cc (30.37cu. in.) in an engine that produces 170bhp @ 13,000rpm. The engine is fitted with twin Mikuni flat slide carburetors and features computer-controlled ignition. A dry multiplate clutch and six-speed transmission are used and these components are fitted to an aluminum twinspar frame. The whole machine is clad with a carbonfiber Kevlar fairing.

1993 SUZUKI RM250

Typical of early 1990s motocross bikes is the Suzuki RM250. The company has had considerable success in off-road competition such as motocross since 1970. The Suzuki RM250 is supplied race-ready and consists of a water-cooled 250cc engine combined with what is described as 'Automatic Exhaust Timing Control' in a tubular steel frame. The whole machine is light in weight at 216lb (98kg) and despite long travel suspension the motorcycle has a seat height of 37.75in. (960mm).

SPECIFICATION
Country of origin: JAPAN
Capacity: 249cc (15.18cu. in.)
Engine cycle: 2-stroke
Number of cylinders: 1
Top speed: n/a
Power: 53.5bhp @ 8500rpm
Transmission: 5-speed
Frame: Tubular steel

1993 SUZUKI RGV250

The RGV250 was Suzuki taking its two-stroke race technology on to the street in a modern sports bike form. The street RGV featured a full fairing behind which was a liquid-cooled, two-stroke V-twin engine. The frame and attached components were equally modern: upside-down forks, alloy wheels, twin disc brakes and at the rear an alloy banana swingarm with disc brake alloy rear wheel.

SPECIFICATION
Country of origin: JAPAN
Capacity: 249cc (15.18cu. in.)
Engine cycle: 2-stroke
Number of cylinders: 2
Top speed: 130mph (209kph)
Power: 62bhp @ 11,000rpm
Transmission: 6-speed
Frame: Aluminum double cradle

SWM

SWM – Speedy Working Motors – started in Milan, Italy, in 1971 and specialized in off-road machines. The company used Sachs and then Rotax engines ranging from 50cc to 320cc (3.20 to 19.52cu. in.). The company took over Gori in 1979 but it went out of business in 1985. An attempt to resurrect the company under the name SVM lasted only between 1985 and 1987. In 1981 a SWM ridden by Giles Burgat won the World Trials Championship.

1980 SWM RS250 GS

This off-road motorcycle was one of SWM's comprehensive range of trials and motocross bikes which was supported by the manufacture of step-through mopeds and similar products. The RS250 GS was also made available with a smaller engine as the RS175 GS.

SPECIFICATION
Country of origin: ITALY
Capacity: 247cc (15.06cu. in.)
Engine cycle: 2-stroke
Number of cylinders: 1
Top speed: n/a
Power: 41bhp @ 8250rpm
Transmission: 6-speed
Frame: Steel cradle

T

TERROT

Terrot, founded in 1901 in Dijon, France, by Charles Terrot, grew into a major motorcycle manufacturer. Proprietary engines from Bruneaux, Dufaux, JAP, MAG and Zedel were utilized at various times. During the 1920s Terrot acquired Magnat Debon, another French manufacturer, but themselves became part of Peugeot in the 1950s. Production ceased in 1961.

1924 TERROT 175

This lightweight machine took advantage of the French taxation on motor vehicles during the 1920s that made small capacity machines popular. It was typical of its time with the fuel tank between the two top tubes of the diamond frame – it was clearly bicycle-inspired. Although conventional in its overall design, the Terrot 175 featured unusual cooling fins on the inclined cylinder barrel.

SPECIFICATION
Country of origin: FRANCE
Capacity: 173cc (10.53cu. in.)
Engine cycle: 2-stroke
Number of cylinders: 1
Top speed: 30mph (50kph)
Power: 3bhp
Transmission: 2-speed
Frame: Tubular diamond

THOR

Thor was the brandname used by The Aurora Automatic Machine Company in the USA. The company manufactured engine castings which were designed by Indian's Oscar Hedstrom for the Springfield marque. The castings were also supplied to other American manufacturers of the time including Emblem, Racycle and Reading Standard. Between 1907 and 1919 complete motorcycles were built under the Thor name.

TOMOS

Tomos is an acronym of Tovarna Motornih Vozil and is a manufacturer based in Koper in the former Yugoslavia. Founded in 1956 it has concentrated on 49cc (2.98cu. in.) machines including step-through mopeds. The company exports widely.

1986 TOMOS SILVER BULLET A3SP

The A3SP is one of the vast Tomos range of small capacity machines that also includes variants styled as sports and trail bikes, such as the BT50 and ATX50 respectively.

SPECIFICATION
Country of origin: FORMER YUGOSLAVIA
Capacity: 49cc (2.98cu. in.)
Engine cycle: 2-stroke
Number of cylinders: 1
Top speed: 30mph (50kph)
Power: 1.8bhp @ 5500rpm
Transmission: 2-speed
Frame: Step-through

TRIUMPH

The Triumph marque is one of Britain's most famous and one with a long and distinguished history. The company was founded in Coventry, England, by two Germans, Siegfried Beltman and Maurice Schultze, in 1897. They initially produced bicycles but by 1902 had made a motorcycle powered by a 220cc (13.42cu. in.) Minerva engine. Later they used Fafnir and JAP engines until Schultze designed the company's first engine. It was of a side-valve design and developed 3.5hp. It was manufactured in two displacements of 499cc and 547cc (30.43 and 33.36cu. in.).

By the 1920s Triumph were producing a variety of machines and co-operated fully with the German Triumph works in Nuremburg until 1929 after which German Triumphs were known as TWN - Triumph Werke Nurnberg. The first parallel vertical-twin from Triumph was designed by Val Page and announced in 1933. It was made until 1936. In that year Triumph's car and motorcycle divisions went separate ways. Jack Sangster of Ariel bought Triumph and put Edward Turner in charge of the Coventry factory. Turner designed the overhead-valve parallel-twin of 498cc (30.43cu. in.) which appeared in the Speed Twin of 1937. This engine would form the basis of

BELOW: The Triumph Tiger of 1937 was powered by an overhead-valve, single-cylinder engine that displaced 493cc (30.07cu. in.) through a bore and stroke of 84 and 99mm respectively. It produced 28bhp @ 5800rpm.

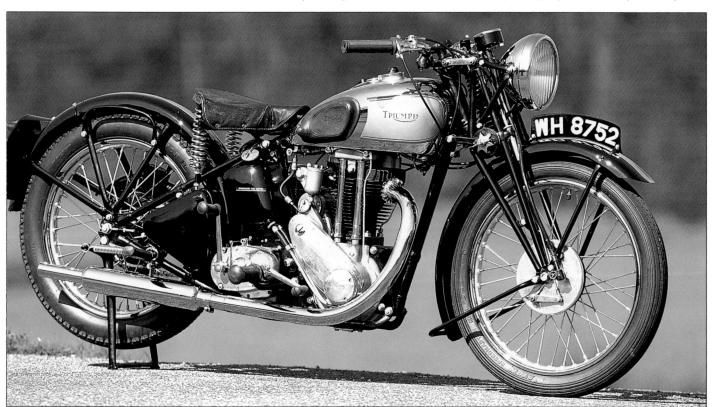

RIGHT: The production of Triumph motorcycles was interrupted by World War II as the company was blitzed out of its Coventry factory. The 1949 Thunderbird was one of the postwar bikes built for export to the USA.

many Triumph motorcycles until the 1980s in bikes such as the Daytona, Bonneville, Tiger and Thunderbird. The Triumph company were blitzed out of their Coventry factory in 1940 and moved to Meriden where their factory was to become famous. Production was resumed in 1944. Postwar Triumph had numerous racing successes including wins at events such as the prestigious Daytona 200 which gave rise to a special model.

During the 1970s the British motorcycle industry entered a period of decline that saw many companies merge or cease production. Triumph became part of a group that comprised BSA, Norton–Villiers and Triumph. These companies all produced Nortons in Wolverhampton, Triumphs at Meriden and BSA supplied engines for the famous Triumph Trident model from the Smallheath plant. The contraction continued and in 1973 the management made the decision to close the Meriden plant. This decade of difficult labor relations was about to see an unprecedented move in the engineering industry. The Meriden workforce occupied the plant and staged a sit-in. The sit-in lasted 18 months and during this time the British Government, the workers representatives and representatives of NVT, as the group was known, negotiated. The result was that a co-operative was formed to produce Triumphs while NVT went on to produce the Norton Commando. Sadly, the co-operative failed after struggling valiantly, due to being desperately short of funds

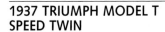

BELOW: The TR100R, a 500cc unit twin, Triumph Daytona was launched in 1967, following the company's victory at the prestigious Florida race venue. The motorcycle was styled especially for the US market and it remained in production until 1973. This is the 1971 model. One of its major selling points was its twin carburetors.

with which to develop new models. Briefly Triumph Bonneville production was shifted to Devon, England, where Les Harris built limited numbers of machines under license from the new owners of the trademark.

John Bloor, a Midlands industrialist, had bought the marque from the receivers and developed a new range of Triumph motorcycles in a new factory in Hinkley, England. These bikes were very popular and re-established Triumph as a premier marque.

1937 TRIUMPH MODEL T SPEED TWIN

This was the first production motorcycle to use Edward Turner's parallel-twin engine. It used cycle parts from Triumph's range of singles and the bike was supplied finished in Amaranth Red. It was immediately popular and it started the trend towards vertical-twins. Production was interrupted by World War II.

SPECIFICATION
Country of origin: GREAT BRITAIN
Engine cycle: 4-stroke
Number of cylinders: 2
Top speed: 93mph (149kph)
Power: 27bhp @ 6300rpm
Transmission: 4-speed
Frame: Tubular steel

ABOVE: The 1937 Speed Twin was smooth and powerful due to its two cylinders of 63 and 80mm bore and stroke. It was designed so that one piston was on the exhaust stroke as the other piston was on the induction stroke.

1940 TRIUMPH 3HW

On the outbreak of war in 1939 the British military quickly standardized its motorcycle requirements and took a single model type from each British factory. Prior to Triumph's Coventry works being destroyed in the blitz, it supplied numbers of the 3HW to the British Army. The 3HW was a military version of the prewar Tiger 80 single.

SPECIFICATION
Country of origin: GREAT BRITAIN
Capacity: 343cc (20.92cu. in.)
Engine cycle: 4-stroke
Number of cylinders: 1
Top speed: 70mph (112kph)
Power: 17bhp @ 5200rpm
Transmission: 4-speed
Frame: Steel tubular

RIGHT: While the 3HW of 1940 was a single built for the British Army, the 3TW (shown) was an overhead-valve twin which was being developed. Due to destruction of the factory by the Luftwaffe, it never went into production.

1971 TRIUMPH DAYTONA T100R

Triumph's numerous race victories and top three finishes during the early 1960s in the Daytona 200 led to their marketing a twin carburetor version of the 500cc parallel-twin, named the Daytona. It used a unit construction engine and gearbox as did the T100C Trophy.

SPECIFICATION
Country of origin: GREAT BRITAIN
Capacity: 489cc (29.82cu. in.)
Engine cycle: 4-stroke
Number of cylinders: 2
Top speed: n/a
Power: 27bhp @ 6500rpm
Transmission: 4-speed
Frame: Steel tubular

1974 TRIUMPH T160 TRIDENT

The Triumph Trident triple appeared in 1968. This motorcycle was derived from the 500cc twin with an extra cylinder added. It became the basis of the BSA Rocket Three and the Triumph Trident and once again scored Daytona 200 race victories. The final version of the triple was the T160 of 1974. This machine was to remain in production until 1976. As a model, it was longer, lower and sleeker than its predecessor and the frame was derived from the Slippery Sam racers.

ABOVE: The Triumph Trident was a triple cylinder machine built in two guises: the European (shown) and the American specification, as well as a custom version.

SPECIFICATION
Country of origin: GREAT BRITAIN
Capacity: 740cc (45.14cu. in.)
Engine cycle: 4-stroke
Number of cylinders: 3
Top speed: 119.5mph (192kph)
Power: 58bhp @ 7250rpm
Transmission: 5-speed
Frame: Duplex tube cradle

1974 TRIUMPH T140 BONNEVILLE

The Triumph Bonneville was enlarged to 720cc (43.92cu. in.) displacement in 1973, the company having already redesigned the frame to act as the oil tank. These bikes are referred to as oil-in-frame Bonnevilles to differentiate them from the earlier models. The machine was available in European and

BELOW: The Triumph Bonneville was launched in 1959, its name reflecting the speed records set at the Utah salt flat on Triumph twins in the 1950s. This is a 1978 model.

American specifications. In this American specification, the new model was described by the magazine *Cycle World* as 'the best Bonneville to date'. It featured a disc front brake and conical rear hub drum.

SPECIFICATION
Country of origin: GREAT BRITAIN
Capacity: 720cc (43.92cu. in.)
Engine cycle: 4-stroke
Number of cylinders: 2
Top speed: 107mph (172kph)
Power: n/a
Transmission: 5-speed
Frame: Tubular steel

ABOVE: One aspect of John Bloor's Triumph operation has been to give the new Triumphs the old model names. This Tiger is an enduro bike based on the liquid-cooled triple engine.

1977 TRIUMPH TR7RV TIGER

The TR7 was a single carburetor version of the Bonneville based around the same displacement parallel vertical-twin. In 1977 it was built by the motorcycle co-operative that was operating in the Meriden factory against the odds. The workers all earned a standard wage and kept the factory going simply because of the demand for Triumphs in the US market which accounted for 70 percent of sales.

SPECIFICATION
Country of origin: GREAT BRITAIN
Capacity: 744cc (45.38cu. in.)
Engine cycle: 4-stroke
Number of cylinders: 2
Top speed: 112.4mph (180kph)
Power: 46bhp @ 6200rpm
Transmission: 5-speed
Frame: Duplex tube cradle

1985 TRIUMPH BONNEVILLE

This motorcycle was one of those machines which was produced by Les Harris under license in Newton Abbot, Devon, England. It was manufactured in two specifications: USA and European. The main differences were in the styling of the seat and tank and in the type of handlebars fitted. The motorcycle was a slightly refined version of the long running 750 Bonneville from Meriden.

SPECIFICATION
Country of origin: GREAT BRITAIN
Capacity: 744cc (45.38cu. in).
Engine cycle: 4-stroke
Number of cylinders: 2
Top speed: n/a
Power: n/a
Transmission: 5-speed
Frame: Steel tubular cradle

LEFT: The first Bonnevilles, such as this 1970 model, displaced 650cc (39.65cu. in.), but this capacity was increased due to US demand.

1991 TRIUMPH TRIDENT 900

The new model Triumphs used names that had been made famous by the original Triumph concern but which were in keeping with the image. The new Trident was based on a triple-cylinder engine and was available in a 750cc (45.75cu. in.) displacement as the original Trident had been. It was also

ABOVE: The Trident 900 of 1991 was, like its famous predecessor, based around a three-cylinder engine but the new model was liquid-cooled and its cycle parts of 1990s standard.

BELOW: The Speed Triple is a mixture of old and new; timeless black paintwork and a clever play on an old name but with a modern three-cylinder liquid-cooled engine.

made in a displacement of the 900cc class. Apart from being a much more modern bike than the original, the new Triumph differed in that the engine was liquid-cooled. It featured three Mikuni carburetors, digital ignition and electric start. The remainder of the machine was technologically up to date and featured cast wheels which were 17in. front and 18in. rear (425 and 450mm), disc brakes front and rear, a monoshock rear suspension system, and adjustable forks.

SPECIFICATION
Country of origin: GREAT BRITAIN
Capacity: 885cc (53.98cu. in.)
Engine cycle: 4-stroke
Number of cylinders: 3
Top speed: 136mph (218kph)
Power: 100bhp
Transmission: 6-speed
Frame: High tensile steel spine

1991 TRIUMPH TROPHY 1200

The Trophy was a sports bike in the style of the Japanese superbikes of the era. It featured a high performance, liquid-cooled, four-cylinder engine with full fairing as well as alloy wheels, disc brakes and an alloy swinging arm.

SPECIFICATION
Country of origin: GREAT BRITAIN
Capacity: 1180cc (71.98cu. in.)
Engine cycle: 4-stroke
Number of cylinders: 4
Top speed: 149mph (240kph)
Power: 125bhp @ 9000rpm
Transmission: 6-speed
Frame: Steel tubular spine

1993 TRIUMPH SPEED TRIPLE

The name of this bike is derived from the old time Triumph Speed Twin but adapted to suit the triple-cylinder engine used in this motorcycle's construction. The Speed Triple was designed as a sports roadster and has found its place on racetracks in a series designed especially for it.

SPECIFICATION
Country of origin: GREAT BRITAIN
Capacity: 885cc (53.98cu. in.)
Engine cycle: 4-stroke
Number of cylinders: 3
Top speed: n/a
Power: 98bhp
Transmission: 6-speed
Frame: Steel spine

LEFT: The Triumph Trophy is available in two displacements but rather than increasing the bore and stroke for the larger model, it has a four-cylinder engine while the smaller machine has a three-cylinder.

LEFT: The Thunderbird is another of the new Triumphs that combines old and new; it features the liquid-cooled, three-cylinder engine but overall the motorcycle is distinctly traditional in appearance.

1996 TRIUMPH ADVENTURER 900

Triumph followed the success of their nostalgic Thunderbird with another machine in a similar style. The Adventurer is more of a cruiser motorcycle even though it is based around the same liquid-cooled, three-cylinder engine, and monoshock frame. The Triumph Adventurer has high pullback handlebars, a nostalgic set of tank badges and a solo seat over a custom-styled rear fender. Engine parts are chromed and the exhaust pipes are the English megaphones.

SPECIFICATION
Country of origin: GREAT BRITAIN
Capacity: 885cc (53.98cu. in.)
Engine cycle: 4-stroke
Number of cylinders: 3
Top speed: n/a
Power: 74bhp
Transmission: 5-speed
Frame: Steel monoshock

BELOW: The 1996 Triumph Adventurer follows the success of the nostalgic Thunderbird. It is a cruiser motorcycle with custom touches including two-tone paintwork, high handlebars and custom fender. It is seen here with the optional dual seat and back rest.

1995 TRIUMPH THUNDERBIRD

The 'new' Thunderbird, as it is known, is a retro-styled bike that conjures up images of the old parallel-twin powered machine but uses a version of the three-cylinder, water-cooled Triumph engine. The Thunderbird was popular and suited the 1990s' trend towards retro-bikes.

SPECIFICATION
Country of origin: GREAT BRITAIN
Capacity: 885cc (53.98cu. in.)
Engine cycle: 4-stroke
Number of cylinders: 3
Top speed: n/a
Power: 70 PS @ 8000rpm
Transmission: 5-speed
Frame: Steel monoshock

ABOVE: The 1995 Triumph Thunderbird is a liquid-cooled, three-cylinder motorcycle, styled traditionally and named after the original Triumph Thunderbird. The modern Triumphs are built by John Bloor's English company.

V

VAN VEEN

Van Veen Import bv was the sole Dutch importer for German Kreidler mopeds and sought to produce a fast sports bike. The company already had experience in developing Kreidler mopeds for small capacity racing classes. Henk Van Veen was the man

BELOW: Henk Van Veen's Dutch-based company developed imported German Kriedler mopeds for racing before going on to build the famous, fast and expensive OCR in the 1970s.

behind the project and chose a rotary engine for the project that commenced in January 1972. The motorcycle he ultimately produced was the OCR 1000 of 1977.

1977 VAN VEEN OCR 1000

This motorcycle was constructed around the Comotor, a rotary engine developed by Audi/NSU with Citroën of France. The bike incorporated expensive quality components throughout including Brembo brakes, Koni suspension components front and rear, a Porsche-developed gearbox, shaft drive and Bosch transistorized ignition. It also used stainless steel for component parts such as the exhausts.

SPECIFICATION
Country of origin: HOLLAND
Capacity: 996cc (60.75cu. in.)
Engine cycle: Rotary
Number of cylinders: Twin rotors
Top speed: 125mph (201kph)
Power: 100bhp @ 6000rpm
Transmssion: 4-speed
Frame: Steel tubular cradle

VELOCETTE

The company, known as Taylor–Gue, a cycle manufacturer, started out by manufacturing frames for another concern which folded. Taylor–Gue then started production of their own motorcycles under the Veloce name in 1905. By 1910 a 276cc (16.83cu. in.) four-stroke motorcycle was in production. This machine was advanced for its day featuring as it did wet sump lubrication and a unit construction engine and gearbox although it was only two-speed. In 1913 a two-stroke model of 206cc (12.56cu. in.) displacement was introduced and known as the Velocette, a name which was to be used on all subsequent machines.

The Model K motorcycle appeared in the years between the two World Wars and was the first Velocette with an engine designed by one of the Goodman family. Subsequently three generations of this family were to control the firm. Through the late 1920s a number of Isle of Man TT vic-

BELOW: The Velocette KSS Mk 1 was intended as a sports road bike version of the KTT. The KSS featured a 348cc (21.22cu. in.) engine and a three-speed gearbox.

tories had gone to Velocette who were using the Goodman overhead camshaft engine. This engine of 348cc (21.22cu. in.) displacement was developed into a range of models, namely the KSS, KTT and KTP, and these models remained in slightly changed form as the basis of Velocette's range along with the 248cc (15.12cu. in.) MOV for several years. The MSS, a 495cc (30.09cu. in.) model was introduced in June 1935. Throughout the years of World War II the company manufactured products other than motorcycles as well as a number of motorcycles for the British war effort.

Production of the company's black and gold machines resumed after the war with a range similar to that of 1939. The real news came in 1949 with the introduction of the LE Model. It was powered by a 149cc (9.08cu. in.) horizontally-opposed, side-valve twin engine with water-cooling. The machine also featured a three-speed, hand-change gearbox and extensive weather protection. The LE was an attempt to capture the mass market in basic transportation in the manner of the Italian scooter manu-

facturers although the Velocette never seemed to quite do so. By 1951 the Velocette range consisted only of the MAC and the LE. The MAC was a traditional overhead-valve, single-cylinder motorcycle and there were two versions of it for 1953. The MSS was reintroduced in revised form and later in scrambler trim while the LE was upgraded by fitting larger diameter wheels.

In the middle of the 1950s Velocette introduced the Viper and Venom models, of 349 and 499cc (21.28and 30.43cu. in.) displacement respectively. Both were high performance sports bikes. These were followed by others including the Viper and Venom Clubman and later still by the Thruxton Venom, named after the famous English race circuit. The Thruxton was introduced in 1965 and in many ways was Velocette's swansong as the company closed in 1971.

1929 VELOCETTE KTT

The KTT model was one of a range of three machines with similar designations based on the 1928 works race model of 1928. The

ABOVE: **The 1929 KTT was almost a race bike and was sold for either racing or road use. It was based around the racers that had brought Velocette three Junior Isle of man TT wins.**

Mark I KTT was closest to the racing bike, a production racer on sale to the public. The KSS was similar and intended as a sports road bike and finally the KTP which featured coil ignition and a dynamo in place of the magneto. Otherwise the machines were similar using as they did a 348cc (21.23cu. in.) overhead camshaft engine. The range endured into the early 1930s and was gradually refined. The K designation endured much longer.

SPECIFICATION
Country of origin: GREAT BRITAIN
Capacity: 348cc (21.22cu. in.)
Engine cycle: 4-stroke
Number of cylinders: 1
Top speed: 70mph (112kph)
Power: n/a
Transmission: 3-speed
Frame: Tubular steel

1948 VELOCETTE KTT MK VIII

The single-cylinder engine in the KTT models had its origins in the years prior to World War II. However, when the firm recommenced motorcycle production after the war, the basic design concept was such that it remained both competitive and reliable until well into the 1950s. Velocette did not

BELOW: A 1948 Velocette Mk VIII KTT. It was the sole motorcycle in Velocette's range to retain girder forks, the others being updated with Dowty Oleomatic telescopic forks.

really build many military motorcycles during the war. However, the company was involved with high precision machining for other products.

SPECIFICATION
Country of origin: GREAT BRITAIN
Capacity: 348cc (21.22cu. in.)
Engine cycle: 4-stroke
Number of cylinders: 1
Top speed: 115mph (185kph)
Power: 34bhp
Transmission: 4-speed
Frame: Tubular steel cradle

1952 VELOCETTE LE 200

The LE was a radical motorcycle for its day and a remarkable achievement by the Goodman family who possessed considerably less resources than companies such as Triumph and BSA. Unfortunately it was introduced prior to the mass markct success of Italian scooters. The LE had numerous novel features including a monocoque frame, adjustable pivoted fork rear suspension, hand lever starting and a handchange transmission. A horizontally-opposed, twin-cylinder engine with water-cooling and shaft drive was a departure in engineering style for the company too. The LE was introduced in 1949 but by 1952 had been improved with the increasing of its displacement from 149cc (9.08cu. in.) to 192cc (11.71cu. in.).

SPECIFICATION
Country of origin: GREAT BRITAIN
Capacity: 192cc (11.71cu. in.)
Engine cycle: 4-stroke
Number of cylinders: 2
Top speed: 55mph (88kph)
Power: n/a
Transmission: 3-speed
Frame: Monocoque

BELOW: The LE Velocette was a motorcycle which, although unusal in appearance, was a creditable attempt by Velocette to capture a share of the very lucrative mass market. This is a 1952 model which is fitted with a 192cc (11.71cu. in.) engine.

1965 VELOCETTE THRUXTON

This model from the Velocette factory was a tuned version of their Venom and named after the English race in which the marque had success. The race was the Thruxton Nine Hour Production race and the new bike was introduced in the year the event was moved to another English circuit. Despite this the Velocette machines dominated in the 500cc class, thereby enhancing the Thruxton's reputation.

SPECIFICATION
Country of origin: GREAT BRITAIN
Capacity: 499cc (30.43cu. in.)
Engine cycle: 4-stroke
Number of cylinders: 1
Top speed: 105mph (170kph)
Power: 40bhp @ 6200rpm
Transmission: 4-speed
Frame: Tubular steel cradle

BELOW: A 1961 Velocette Venom. It became the Thruxton following the tradition of naming motorcycles after places where they had enjoyed competition success, as Venoms had in the Thruxton Nine Hour Production race.

1969 VELOCETTE INDIAN 500

The final days of the Indian company of Springfield, Massachusets, USA, saw them importing and retailing British machines badged as Indians. One such was this 500cc (30.50cu. in.) single-cylinder Velocette and another was a 750cc (45.75cu. in.) twin-cylinder Royal Enfield.

SPECIFICATION
Country of origin: GREAT BRITAIN
Capacity: 500cc (30.50cu. in.)
Engine cycle: 4-stroke
Number of cylinders: 1
Top speed: n/a
Power: n/a
Transmission: 4-speed
Frame: Tubular steel

VESPA

Enrico Piaggio had run a heavy industrial operation that had been founded in 1884 in Pontadera, Italy. It had turned its attention to the aviation industry in 1915. By 1945 the plant was rubble, destroyed during World War II as a direct result of its production of bombers. The management of the company were anxious to find a product for their workforce to build in the immediate postwar period of reconstruction. With American assistance the company looked towards fulfilling the need for basic transportation with a two-wheeler capable of negotiating rough roads. The company had been making a 98cc (5.97cu. in.) engine to drive aeroplane generators so built its first two-wheeled machine using one of these engines. It was a radical design for its time and had been developed by Corradino D'Ascanio. It was his second design that really established the Vespa – which is the Italian for 'Wasp' – as a serious means of transport. Many of its design features reflected the company's aerospace background including stub axles and a stressed skin monocoque. Demand for scooters was massive and immediate and their stylish shape reflected aspirations for the future.

The Vespa had a neatly designed engine and gearbox that were positioned side by side and cooled by a fan. The whole unit moved up and down as part of the suspension system. The engine is acknowledged as one of the most efficient two-stroke power units ever and it was later enlarged and developed rather than redesigned. Vespa scooters were later assembled under license by a British company, Douglas.

LEFT: Although the Vespa scooter was introduced in the years immediately following World War II, its design has remained essentially unchanged, as this 1986 PK50XL shows. It was aimed at younger riders.

1980 VESPA P200E

In 1980 Piaggio was the fourth largest two-wheeler manufacturer in the world and built machines including Gilera (acquired in 1969) motorcycles in several factories. The P200E is typical of the company's scooters of that time: a two-stroke rotary valve, single-cylinder engine under the seat, leg-shields and small wheels.

SPECIFICATION
Country of origin: ITALY
Capacity: 197.97cc (12.01cu. in.)
Engine cycle: 2-stroke
Number of cylinders: 1
Top speed: 68mph (110kph)
Power: 12.35bhp @ 5700rpm
Transmission: 4-speed
Frame: Monocoque

ABOVE: The Vespa P200E was one of a range of machines of similar appearance, available in displacements of 50, 100, 125 and 150cc (3.05, 6.10, 7.60 and 9.15cu. in. respectively) and in three- and four-speed variants.

VICTORIA

This German company was founded in 1886 by Max Frankenburger and Max Ottenstein as a bicycle manufacturing concern. The first motorcycles were produced in 1912 utilizing Fafnir and Zedel proprietary engines. In the years between the two World Wars production was concentrated on 493cc (30.07cu. in.) horizontally-opposed twins using a BMW proprietary engine. The Nuremburg company hired designer Martin Stolle, who had previously designed a new overhead-valve twin for BMW. The engine factory was at Sedlbauer, Germany, and made 498cc (30.37cu. in.) and later 598cc (36.47cu. in.) engines. The next designer employed by the company was Gustav Steinlein, who designed super-charged racing bikes which broke the German speed record in 1924 with a top speed in excess of 100mph (161kph). In the late 1920s the company introduced new single-cylinder motorcycles using Sturmey Archer engines and later Horex license-built engines of the same type. During the 1930s the company made small capacity two-stroke machines and 497cc (30.32cu. in.) twins which were fitted into a triangular pressed steel frame.

In the years following World War II the company recommenced production with small capacity two-stroke machines with one major exception, the Bergmeister, a

ABOVE: The Victoria name was used by a Scottish company who built motorcycles with proprietary engines including units from JAP, Villiers and Blackburne between 1902 and 1926. The factory was in Dennistoun, Glasgow. There was no connection with the German company of the same name.

BELOW: The Victoria KR9 of 1937 was built by the German users of the name Victoria. It was a utilitarian motorcycle assembled from steel pressings; the frame, forks and sidepanels which enclosed the twin-cylinder engine were all pressed. The bike featured legshields and footboards for the rider's comfort.

347cc (21.16cu. in.) transverse V-twin. This was made from 1951 to 1958 and after this date the company concentrated on the production of lightweight machines as part of the Zweirad Union that also included DKW and Express. By 1966 the brandname had disappeared completely.

1937 VICTORIA KR9

Martin Stolle designed a number of completely enclosed, inlet-over-exhaust, parallel twin-cylinder machines for Victoria which involved the fitting of the unit-construction engine into a pressed steel triangular frame with sidepanels that hid the engine from view. The machines also used pressed steel forks and were utilitarian in appearance.

SPECIFICATION
Country of origin: GERMANY
Capacity: 496cc (30.25cu. in.)
Engine cycle: 4-stroke
Number of cylinders: 2
Top speed: 62mph (100kph)
Power: 15PS @ 4300rpm
Transmission: 4-speed
Frame: Pressed steel

1953 VICTORIA BERGMEISTER

This was the first postwar four-stroke motorcycle made by the German company. The design incorporated a rubber-mounted, transverse V-twin engine and shaft final drive. The Bergmeister had telescopic forks and plunger rear suspension. As a result of being costly to develop its production run was short and the firm was forced to switch back to the production of smaller capacity machines.

SPECIFICATION
Country of origin: GERMANY
Capacity: 347cc (21.16cu. in.)
Engine cycle: 4-stroke
Number of cylinders: 2
Top speed: 80mph (128kph)
Power: 21bhp @ 6300rpm
Transmission: 4-speed
Frame: Plunger

VINCENT–HRD

Philip C. Vincent was a graduate from the University of Cambridge, England, where he had studied Mechanical Science. He had a low opinion of many of the features of motorcycles of the era and so decided to build his own. His first machine incorporated a Swiss MAG engine, a Moss gearbox, Webb forks, and rear suspension through a pivoting triangular rear frame section. From this experimental machine Vincent decided to go into motorcycle production and purchased the established HRD name from OK Supreme. The former marque was well known as its founder Howard R. Davies had won and been placed in a number of Isle of Man TT races before and

ABOVE: The Grey Flash was a single-cylinder built by Vincent in 1950. It displaced 500cc (30.50cu. in.) and was a Comet with a tuned engine, no lights or other road-use equipment, and a number of racing components.

after World War I on machines of his own construction.

By 1930 Vincent–HRD were regarded as makers of high quality, high class and hand-built motorcycles. The company initially used proprietary engines from JAP and then from Rudge but later moved on to manufacture their own brand of engine. These included a number of innovative design features and were introduced in 1935. A number of models were listed as singles and the first V-twin was announced in 1937. The V-twin was made from a combination of two Meteor single-cylinders into a single

crankcase. The result was a fast motorcycle with the V-twin engine fitted to a slightly longer than standard single-cylinder frame. The company ceased motorcycle production in late 1939 and turned its factory over to war work. Irving and Vincent redesigned the Vincent Rapide during the war years and reintroduced it in redesigned form in 1945. A faster version, the Black Shadow, was introduced in 1948 and had a top speed in excess of 120mph (193kph). Some single-cylinder models were produced postwar too, including the Meteor and Comet.

BELOW: The 1951 Vincent Comet, a 500cc (30.50cu. in.) single-cylinder machine, was one of only four roadgoing models and the only single offered by Vincent in that year. The other three machines were V-twins and these four remained as the range until 1954.

1952 SERIES C VINCENT BLACK SHADOW

The Black Shadow was a high performance version of the Rapide which itself was no slouch. Series C motorcycles were introduced in 1949 and remained in production until 1954. They were refined examples of the earlier series and by 1952 included Girdraulic forks, which were manufactured by Vincent, combining girders and hydraulic damping. The sports machines also featured hydraulically-damped rear suspension in an unusual arrangement, twin drum brakes, and the potent V-twin engine. At the time of its manufacture the Vincent had the distinction of being the world's fastest production motorcycle.

SPECIFICATION
Country of origin: GREAT BRITAIN
Capacity: 998cc (60.87cu. in.)
Engine cycle: 4-stroke
Number of cylinders: 2
Top speed: 125mph (201kph)
Power: 55bhp @ 5700rpm
Transmission: 4-speed
Frame: Tubular steel

BELOW: Girdraulic forks were also fitted to the Series C Black Shadow, as seen on this 1952 model. The rake and trail was adjustable to make the machine suitable for sidecar use.

During the mid-1950s production of Vincent motorcycles was halted after a period of co-operation with the German NSU company.

1949 VINCENT RAPIDE

The Rapide name was used on a Vincent before World War II and reintroduced on a redesigned motorcycle soon after it. The postwar Vincent–HRD Rapide was correctly advertised as the 'world's fastest standard motorcycle' and went into production in 1946. By 1948 there were Series B and C Rapides, the major difference being in the type of front forks employed.

SPECIFICATION
Country of origin: GREAT BRITAIN
Capacity: 998cc (60.87cu. in.)
Engine cycle: 4-stroke
Number of cylinders: 2
Top speed: 112mph (180kph)
Power: n/a
Transmission: 4-speed
Frame: Tubular steel

ABOVE: The Vincent V-twins – the Rapide and
the Black Shadow – were described as Series C
machines once they were fitted with the
'girdraulic' forks seen here. The Black Shadow
(shown) was a tuned Rapide.

W

WANDERER

This German company was founded in 1902 and became well known as a manufacturer of quality machines. The company produced 327 and 387cc (19.94 and 23.60cu. in.) singles and side-valve V-twins of 408 and 616cc (24.88 and 37.57cu. in.) displacement with engines of their own design and manufacture. The company supplied a number of its motorcycles to the German Army in World War I. After the armistice the company produced a novel 184cc (11.22cu. in.) machine in which the single overhead-valve cylinder was horizontal. They also manufactured larger displacement V-twins. Toward the end of the 1920s a new machine designed by Alexander Novikoff went into production. It consisted of a 498cc (30.37cu. in.) single-cylinder engine of unit design and shaft-driven rear wheel fitted into a pressed steel frame. The production run was short and in 1929 the whole design and production equipment was sold off to the Janacek company of Prague, Czechoslovakia. Perhaps ironically, the Janacek Wanderer gave a name to the company – Jawa. This sale signified the end of Wanderer bike production although later they manufactured motorized bicycles in conjunction with NSU.

WERNER

This French company was founded by two Russian brothers – Michel and Eugene Werner – who lived in Paris, France. The brothers were amongst the pioneers of the motorcycle and started manufacture in 1897. Their first two-wheeled machines were powered by a Labitte engine that was installed above the front wheel of a bicycle-type machine but later ones had the engine fitted in behind the seat down-tube. The early Werner motorcycles displaced 217cc (13.23cu. in.) and the later ones 230cc (14.03cu. in.). An English pioneer named Lawson founded a Werner factory in Coventry, England, in 1899 and other companies, including one in Germany and one in Austria, purchased Werner patents in order to produce motorcycles. The brothers went on to develop a twin-cylinder engine in 1905 but production ended in 1908 when Michel died.

BELOW: The TMC designation given to the two-wheeled Wilkinson machines stood for 'Touring Motor Cycle'. Prior to this it had been known as the TAC ('Touring Auto Car') as the machine was fitted with a steering wheel. The TMC featured an in-line four-cylinder engine.

WILKINSON

The company that became famous as a sword manufacturer, and later for making razor blades, briefly built an unusual motorcycle. Initially it was called the Wilkinson TAC. This was an acronym for 'Touring Auto Car' and at the time this unusual two-wheeler featured a steering wheel. Later the steering wheel was superseded by handlebars and the machine became known as the TMC – 'Touring Motor Cycle'. The machine was unorthodox, featuring a bucket seat and an inlet-over-exhaust, air-cooled, four-cylinder engine. Initially this unit displaced 676cc (41.23cu. in.), a capacity that was enlarged to 844cc (51.48cu. in.) later.

WILLIAMSON

This was an English manufacturer based in Coventry in the British Midlands who were in business for only eight years between 1912 and 1920. The company had its engines – flat-twins – manufactured by Douglas. They displaced 964cc (58.80cu. in.) and they were available as both water- and air-cooled units. In the years which followed World War I the company used JAP side-valve V-twin 770cc (46.97cu. in.) proprietary engines. The concern was named after Billy Williamson who had founded the company.

ABOVE: The first Williamson machines had side-valve flat-twin engines. This 1913 model has the liquid-cooled variant of this engine.

WINDHOFF

This German company was based in Berlin and came into motorcycle manufacture from the production of radiators. They started motorcycle manufacture in 1924 and produced water-cooled two-strokes of 122 and 173cc.(7.44 and 10.55cu. in.). The engines were horizontal but of a two-piston design. The machines had some success in races. The company went on to produce frameless multi-cylindered machines including their liquid-cooled fours and flat-twins. Before the company closed in 1933 it manufactured conventional two-strokes using Villiers 198 and 298cc (12.07 and 18.17cu. in.) two-stroke engines.

1929 WINDHOFF VIERZYLINDER

This machine appeared in 1928 and consisted of an oil-cooled, overhead camshaft, in-line four-cylinder engine to which all the other parts of the machine were bolted. It did not have a frame but utilized shaft drive and girder forks.

SPECIFICATION
Country of origin: GERMANY
Capacity: 746cc (45.50cu. in.)
Engine cycle: 4-stroke
Number of cylinders: 4
Top speed: 75mph (120kph)
Power: 22PS @ 4000rpm
Transmission: 3-speed
Frame: Frameless

BELOW: The radiator manufacturer's inspiration is obvious in this oil-cooled, in-line four-cylinder engine fitted to the 1929 model Vierzylinder, manufactured by the Berlin-based Windhoff company.

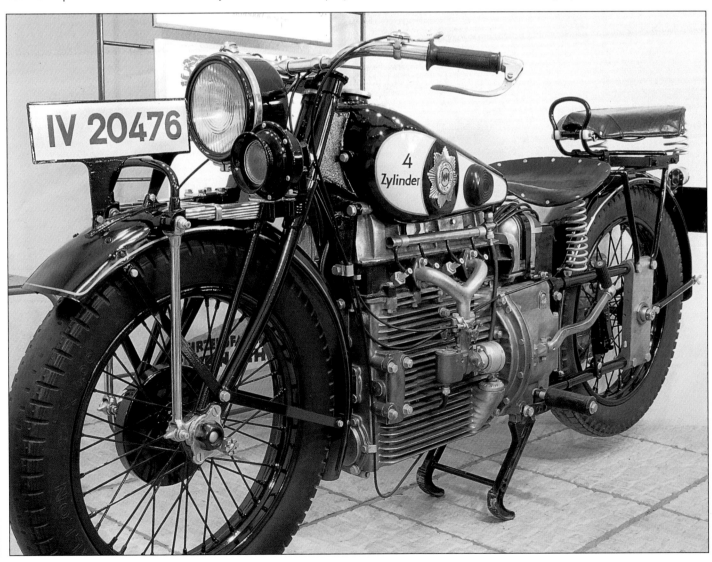

Y

YALE

The Consolidated Manufacturing Company of Toledo, Ohio, began motorcycle manufacture on acquiring an extant California-based company, Yale. From this start in 1902 it went on to produce both singles and V-twin engine designs of motorcycle. The single of 1910 with its 500cc (30.50cu. in.) displacement, relied on an atmospheric inlet valve and battery and coil ignition. The machine was ruggedly engineered in order to withstand travel on the poor roads of the time. In 1910 the company also introduced a 45° V-twin of its own design. The new engine displaced 1000cc (61cu. in.) and had mechanical valve actuation. The 1910 V-twin featured a two-speed transmission and also a clutch mechanism. Production stopped in 1915.

YAMAHA

Yamaha were a musical instrument manufacturer that had been founded in 1887 but diversified into motorcycle manufacture in 1955. Their first machine, the YA1, was a 125cc (7.62cu. in.) two-stroke based closely on the prewar DKW RT125 which was soon followed by 175 and 250cc (10.67 and 15.25cu. in.) models. A new factory was constructed at Iwata, southwest of Tokyo, Japan, and motorcycle production continued apace. Soon the range included 175 and 250cc (10.67 and 15.25cu. in.) models in duplex cradle frames, as well as a 173cc (10.55cu. in.) scooter and a moped. By

ABOVE: **Numerous manufacturers have taken the scooter and step-through concepts and refined them. This is the 1995 Yamaha YE50 Zest, a basic commuter machine.**

1960 Yamaha were exporting motorcycles to the USA and developing specialized race bikes. In 1961 the company entered machines in its first Isle of Man TT race and introduced the first use of a rotary valve engine into production machines with the YA5. Reed valves and automatic lubrication were introduced on 1964 models. The YAS1 of 1967 had a five-valve head to improve engine efficiency. By the end of the 1960s

riders on Yamaha machines had won five World Championships.

In 1969 the Yamaha company introduced its first four-stroke, the XS1. This machine featured a 654cc (39.89cu. in.) overheadcamshaft, parallel twin-cylinder engine and was followed by a range of other displacements. The reason for this was that Yamaha were moving away from its total commitment to two-strokes. As a result their products diversified, although not exclusively toward two-stroke dirt bikes and four-stroke road bikes. A range of large capacity four-strokes appeared in 1972 and by 1977 the company had developed a new generation of four-strokes including the XS750 triple with shaft drive. Later came larger models such as the XS1100. The company's first four-stroke V-twin was unveiled in 1982. Yamaha's two-stroke technology reached its zenith in the mid-1980s when the roadgoing 499cc (30.43cu. in.) V4 was marketed. Increasingly tight emissions legislation and accelerating four-stroke technology meant that the company would in future concentrate its efforts on four-stroke sports bikes.

Yamaha produces in excess of a quarter of Japan's total annual motorcycle output as well as musical instruments, snowmobiles, boats and jetskis. It is part of the Nippon Gakki group.

LEFT: **The Yamaha XJ600S Diversion of 1995 was a sports tourer with a 60bhp double overhead camshaft, transverse four-cylinder engine of 598cc (36.47cu. in.) and capable of 112mph (180kph). This style of bike evolved out of the Paris-Dakar styled off-road bikes of the early 1990s and became very popular.**

1977 YAMAHA XS750

The XS750 was a motorcycle which was traditional in appearance but nonetheless rather unusual for its time. It was a double overhead camshaft triple with shaft drive, cast alloy wheels and disc brakes. It was aimed at touring rather than sporting riders.

SPECIFICATION
Country of origin: JAPAN
Capacity: 747cc (45.56cu. in.)
Engine cycle: 4-stroke
Number of cylinders: 3
Top speed: 117mph (188kph)
Power: 64bhp @ 7200rpm
Transmission: 5-speed
Frame: Duplex cradle

1980 YAMAHA XS650SE

The XS650 had a parallel-twin engine and was a motorcycle not unlike many British bikes of the 1960s and 1970s in appearance and design. It was originally introduced in

ABOVE: The 1980 Yamaha XS650SE was a special version of the XS650 parallel-twin. It featured US custom styling with stepped seat.

ABOVE: The XS650 Yamaha twin came in many styles including factory customs. The Heritage Special of the late 1970s was one such model, featuring a stepped seat and high bars.

1969 as the XS1 and was sequentially upgraded. The brakes were upgraded from drums to discs, an electric starter was fitted, the wheels were changed from spokes to cast alloy items and so on. By 1980 the fad for custom-styled bikes had arrived and the XS650SE was introduced. This model featured the same reliable parallel-twin engine but a number of custom touches including a slightly stepped seat, alloy wheels, pull-back handlebars and a number of other detail changes.

SPECIFICATION
Country of origin: JAPAN
Capacity: 653cc (39.83cu. in.)
Engine cycle: 4-stroke
Number of cylinders: 2
Top speed: 102mph (165kph)
Power: 54.8bhp @ 7000rpm
Transmission: 5-speed
Frame: Duplex cradle

1985 YAMAHA V-MAX

The Yamaha V-Max was a brutal-looking motorcycle, clearly American-influenced in its design and based around the liquid-cooled V4 engine from Yamaha's tourer, the Venture Royale. The V-Max was a factory hot-rod loaded with horsepower and it soon attracted enthusiasts. The motorcycle had 145bhp in standard form but was sold in some markets, including Great Britain, restricted to 95bhp. The V-Max features shaft drive, disc brakes and alloy wheels.

SPECIFICATION
Country of origin: JAPAN
Capacity: 1198cc (73.07cu. in.)
Engine cycle: 4-stroke
Number of cylinders: 4
Top speed: 144mph (230kph)
Power: 145bhp @ 8000rpm
Transmission: 5-speed
Frame: Duplex cradle

BELOW: The 1985 Yamaha V-Max was an awesome and brutal-looking bike, loaded with horsepower. It was renowned for being fast but also for not handling particularly well.

ABOVE: Yamaha manufacture BW's Bi-Wizz, which is a commuter step-through machine. It utilizes a single-cylinder, two-stroke engine, concealed behind the bodywork. Such models are a development of the scooter concept.

ABOVE: The Yamaha TDM was introduced as a sports tourer in 1991 and, with refinements, was still powered by a vertical-twin engine in 1996. This is a 76bhp @ 7000rpm 1995 model.

rear suspension set-up and disc brakes both front and rear. The styling featured a mini-fairing that flowed into the lines of the seat and tank.

SPECIFICATION
Country of origin: JAPAN
Capacity: 849cc (51.78cu. in.)
Engine cycle: 4-stroke
Number of cylinders: 4
Top speed: 136mph (220kph)
Power: 77bhp @ 7500rpm
Transmission: 5-speed
Frame: Steel delta-box

1993 YAMAHA GTS1000

The Yamaha GTS1000 was an advanced motorcycle that displayed a number of interesting design features including hub center steering incorporated into a single-sided front swinging arm suspension assembly. Rear suspension was a monoshock and the two assemblies were connected to what Yamaha called the 'Omega' chassis. Much of this was concealed behind bodywork that was integral with the fairing. Despite the innovative engineering that went into its construction, the hub steer tourer did not sell in huge numbers, possibly as a result of its unorthodox appearance.

SPECIFICATION
Country of origin: JAPAN
Capacity: 1002cc (61.12cu. in.)
Engine cycle: 4-stroke
Number of cylinders: 4
Top speed: 132mph (213kph)
Power: 100bhp @ 9000rpm
Transmission: 5-speed
Frame: Alloy box section

BELOW: The Yamaha GTS1000 was introduced in 1993, the year this example was made. It looked unusual as a result of its hub center steering which gave the motorcycle a poor turning circle.

1995 YAMAHA XT600E

The XT600E is one of Yamaha's large capacity trail bikes with monoshock rear suspension, long travel telescopic forks and disc brakes. Its engine is a single-cylinder unit and available in both 350 and 600cc displacements. The 1995 model features both high ground clearance and contemporary styling.

BELOW: The XT600E is a four-stroke, single-cylinder, air-cooled trail bike designed for both on- and off-road use. It is classic in its design – a big single – and a basic machine under its modern 1995 graphics and paintwork.

SPECIFICATION
Country of origin: JAPAN
Capacity: 595cc (36.29cu. in.)
Engine cycle: 4-stroke
Number of cylinders: 1
Top speed : 105mph (170 kph)
Power: 45bhp @ 6500rpm
Transmission: 5-speed
Frame: Tubular cradle

1996 YAMAHA XVZ13A ROYAL STAR

The Yamaha Royal Star is a motorcycle for the developing cruiser market of the 1990s. Its styling is not dissimilar to that of a big twin Harley–Davidson although the power comes from a liquid-cooled, 70°, double overhead camshaft, V-four engine. The engine was derived from the successful V-Max unit although in a lower state of tune. The pillion seat is removable and Yamaha produce a range of accessories for the machine as well as a touring version known as the Tour Classic. The Royal Star has a 16in. (400mm) front wheel and a 15in. (375mm) rear wheel, both of which are cast from alloy. Brakes are twin discs front and single disc rear and final drive is shaft.

SPECIFICATION
Country of origin: JAPAN
Capacity: 1294cc (78.93cu. in.)
Engine cycle: 4-stroke
Number of cylinders: 4
Top speed: n/a
Power: 74bhp @ 4700rpm
Transmission: 5-speed
Frame: Tubular steel

BELOW: The XVZ13A Royal Star is Yamaha's heavyweight motorcycle for the cruiser market and powered by a liquid-cooled, V-four engine derived from other Yamaha V-four models.

Z

ZÜNDAPP

The Zünder und Apparatebau Company was founded by Fritz Neumeyer in 1917 in Nuremburg, Germany, to make fuses for artillery ammunition. After World War I was over, the company began making motorcycles. Its first model, which was a 211cc (12.87cu. in.) two-stroke appeared in 1921. It soon gained a reputation for reliability. More two-strokes were introduced at the end of the 1920s as was a motorcycle using a 498cc (30.37cu. in.) Rudge engine purchased from England. By 1933 the company had manufactured 100,000 motorcycles. They diversified into three-wheelers but also started manufacturing shaft driven four-strokes. A range of machines was made including 398 and 498cc (24.27 and 30.37cu. in.) flat-twins and 598 and 797cc (36.47 and 48.61cu. in.) flat-fours. Almost the entire range including the DB200, K500W, KS600W, K800W and KS750 machines were supplied by Zündapp to the Wehrmacht during World War II.

After the war, the 198cc (12.07cu. in.) two-stroke and 597cc (36.41cu. in.) flat-twin were reintroduced while the company also made sewing machines. In the 1950s the company introduced its line of Bella scooters and moved production to Munich, Germany. By the 1980s Zündapp was Germany's biggest motorcycle manufacturer but it collapsed in 1984. Some of its tooling and equipment was sold to China.

1943 ZÜNDAPP KS750

The German Army of World War II made much use of motorcycles in both solo and sidecar configurations. BMW, NSU, DKW and Zündapp all supplied large numbers of machines for use in all theaters of operations. The flat-twin KS750 from Zündapp was a motorcycle developed especially for the Wehrmacht after the conditions experienced in desert sands and on the Eastern Front proved too tough for the earlier KS600 and K800 models. It was the same story for BMW. Their products featured modifications specific for the country in which they would be used, e.g. foot heaters for the low temperatures of the Soviet Union and extra air filters for the desert terrain of North Africa. The heavyweight shaft drive machines (which were manufactured in too few numbers to affect the outcome of the Wehrmacht's campaigns) had reverse gear and a low ratio gear intended for use in difficult going, as well as a driven sidecar wheel.

SPECIFICATION
Country of origin: GERMANY
Capacity: 750cc (45.75cu. in.)
Engine cycle: 4-stroke
Number of cylinders: 2
Top speed: n/a
Power: 26bhp
Transmission: 4-speed
Frame: Pressed steel

1980 ZÜNDAPP KS175

Typical of a motorcycle intended as basic transporatation is the KS175 two-stroke. The KS175 features a number of sports-type details such as a mini fairing and cast alloy wheels. Disc brakes are standard having in the main superseded drum brakes.

SPECIFICATION
Country of origin: GERMANY
Capacity: 163cc (9.94cu. in.)
Engine cycle: 2-stroke
Number of cylinders: 1
Top speed: 74.5mph (120 kph)
Power: 17bhp @ 7400rpm
Transmission: 5-speed
Frame: Tubular cradle

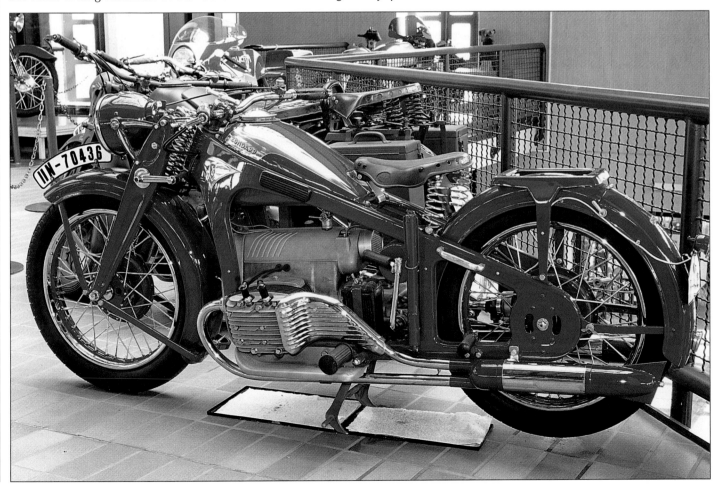

BELOW: A 1930s Zündapp K800. It featured shaft drive, four-speed transmission and a flat-four engine of 800cc (48.80cu. in.). It was renowned for being a luxury sports machine.

GLOSSARY

A

Air cooling
Motorcycle engines without radiators are cooled by air flowing around the cooling fins.

Alloy
A metal created by mixing two or more metals for a particular purpose such as strength, lightness, resistance to corrosion and so on.

Alternator
An electrical generator that makes alternating current by spinning a magnetic rotor inside a coil-wound stator.

B

bhp
Brake horsepower. A unit of measurement for engine power output; see *power*.

Bore
The measurement of a cylinder's diameter; see *stroke*.

C

Cam
An eccentrically-shaped rotor that converts rotational movement into linear movement. Cams are used in the operation of valves, drum brakes and contact breaker points.

Camshaft
A shaft with two or more cams that is used in the four-stroke engine to operate the inlet and/or exhaust valves.

Capacity
The displacement of a motorcycle engine measured in cubic centimetres or cubic inches.

Carburetor
The device that mixes the fuel and air into a combustible vapor.

Chopper
A highly modified custom motorcycle.

Coil
A form of ignition that improves upon the magneto.

Contact breaker
The sprung switch in the low-tension ignition circuit that controls the timing of the spark in the high-tension circuit. Sometimes called points, the contact breakers are operated by a cam.

Country of origin
The country where a particular motorcycle is made.

Cradle frame
A frame that has a cradle around the engine.

Crankshaft
The cranked shaft in an engine that changes the linear motion of the piston into rotational motion.

Customize
The process of bike conversion to the individual's requirements.

Cylinder head
A casting that caps the cylinder and contains the combustion chamber and the valves.

D

Diamond frame
Old bicycle style tubular frame design that was common until World War II. The profile resembles a diamond, hence the name.

Dirt-track
A form of American motorcycle racing.

Distributor
A device that is used in the ignition system of some multi-cylinder machines to send the high-tension spark to the correct cylinder.

Duplex cradle
A cradle frame with twin front downtubes. Duplex also refers to chains with double rows of rollers.

Dynamo
An electric generator that produces alternating current.

E

Engine cycle
An indicator of whether a machine possesses a two-stroke or a four-stroke engine.

Exhaust valve
The valve that allows the burned fuel out of the cylinder.

F

Factory custom
A factory produced motorcycle that appears as though customized.

Fairing
An enclosure that is fitted to improve the areodynamic performance of the machine, although sometimes it is fitted merely to enhance rider comfort.

Final drive
The drive chain or belt from gearbox to rear axle that provides the means of transmitting power to the driven wheel.

Flat-twin
A twin-cylinder engine with horizontal opposing cylinders.

Frame
The type of frame used in the construction of a motorcycle.

Fuel injection
Process whereby fuel is introduced directly under pressure into the combustion unit of the internal combustion engine.

G

Gear ratio
The ratio of the turning speeds of a driving and driven gear.

Girder forks
Sprung forks that work on a parallelogram principle to give front suspension. Common on early machines.

H

Helical gear
A gear with a spiral or semi-spiral meshing face.

Horizontally opposed
An engine layout in which the cylinders are placed at 180° to one another.

I

Inlet valve
The valve that allows the fuel/air mixture into the cylinder.

In-line four
A four-cylinder motorcycle engine with four cylinders along the frame.

GLOSSARY

K

kph
Kilometres per hour.

L

Leaf sprung
A form of suspension consisting of strips of spring steel clamped together with one end fixed and the other attached to the sprung component.

Liquid cooling
Where coolant and a radiator is used to cool a motorcycle engine.

M

Magneto
An electric generator using permanent magnets to create a high-tension spark without the use of an external power source such as a battery.

Monoshock
A single shock absorber rear suspension arrangement.

Moped
A motorcycle of less than 50cc displacement.

Moto-X
A form of off-road motorcycle race formerly known as scrambles and also called motocross.

mph
Miles per hour.

N

Number of cylinders
The number of cylinders in a motorcycle engine.

O

Off-road
A term for trials and motocross environments.

Overhead camshaft
See *camshaft*.

Overhead-valve
Engines with the valves in the cylinder head situated over the pistons and operated by pushrods controlled by a camshaft that is situated below it.

P

Parallel-twin
A two-cylinder engine with the cylinders vertical and parallel to each other.

Power
The horsepower produced by an engine usually measured in bhp (brake horsepower).

Primary Drive
The system of drive or transferance of power from the engine to the transmission.

Pushrod
Metal rod that is used to transmit linear motion; most often this is from the camshaft to the rocker arm; see *overhead valve*.

R

Race-replica
A roadgoing sports bike that looks like a racing bike.

Rigid frame
One without rear suspension.

Road racing
Grands Prix-type race events held on closed public roads.

Rotary engine
An engine that works on the Wankel rotary principle.

S

Saddle tank
A fuel tank that is fitted over the top tube of the bike frame.

Scooter
Small-wheeled utilitarian machine with a step-through frame.

Servo
Any system that is used to assist a mechanism to operate with greater force than that which is initially applied to it.

Side-valve
An engine with valves that are positioned in the side of the combustion chambers.

Speedway
A form of bike racing on a short dirt oval track.

Springer forks
A variation on the girder forks.

Step-through
A specific design of moped frame.

Stroke
Measurement of the length of piston travel in the bore, usually given in millimetres; see *bore*.

Supercharger
A mechanically-powered device that compresses the combustible charge into the cylinder and artificially increases the compression ratio.

Swingarm
A fork that holds the rear axle and pivots to provide rear suspension.

T

Telescopic forks
Forks with tubes that slide into each other to give front suspension, usually hydraulically damped.

Top speed
The top speed in kph and mph attainable by a particular machine.

Transmission
The type of gearbox and number of gears fitted.

Transverse four
An across-the-frame four-cylinder motorcycle engine.

Trials bike
A motorcycle designed for off-road trials competition.

TT
The world famous Tourist Trophy races held on the Isle of Man.

Turbocharger
Similar to the supercharger but the compression is driven by exhaust gas.

V

V-twin
A twin-cylinder engine with the cylinders arranged in a V-position.

INDEX

PICTURE CREDITS

The publishers are grateful to the following institutions and individuals that have provided the photographs in this book.

ABBREVIATIONS
NMMB: The National Motor Museum, Beaulieu
Quadrant: Quadrant Picture Library
CM: *The Classic Motorcycle* Magazine, EMAP Publications
JC: John Carroll
AC: Alan Cathcart
MM: Mac McDiarmid
GS: Garry Stuart
H: Honda
K: Kawasaki
Y: Yamaha

Page 1: GS; **2:** GS; **4:** GS; **8:** (top right) CM, (bottom) GS; **9:** (top left and center) CM, (bottom) MM; **10:** (top left and top right) CM, (bottom) MM; **11:** Quadrant; **12/13:** GS; **14:** (top and bottom) MM, (center) CM; **15:** (top) NMMB, (center) CM, (bottom) GS; **16:** (all) CM; **17:** (top) CM, (center and bottom) NMMB; **18:** (top) Quadrant, (bottom) MM; **19:** (top and bottom) MM, (center) CM; **20/21:** GS; **22:** (top) NMMB, (center and bottom) MM; **23:** (top) MM, (bottom) NMMB; **24:** (top and center) NMMB, (bottom) GS; **25:** (top) GS, (center and bottom) NMMB; **26:** (top) GS, (bottom) NMMB; **27:** (top) NMMB, (center) CM, (bottom) GS; **28:** (top) Quadrant, (bottom) CM; **29:** (all) CM; **30:** (top left and bottom right) NMMB, (top right and bottom left) CM; **31:** (top left and bottom left) GS, (top right and bottom right) NMMB; **32/33:** GS; **34:** (top) CM, (center and bottom) MM; **35:** (all) CM; **36:** (top left) CM, (top right) GS, (bottom) NMMB; **37:** (top) CM, (left) Quadrant, (bottom) GS; **38/39:** GS; **40:** (top) JC, (bottom) MM; **41:** (top) JC, (bottom) NMMB; **42:** (top) NMMB, (bottom) MM; **43:** (both) CM; **44:** (top left and bottom) CM, (top right) NMMB; **45:** (top left) NMMB, (top right, bottom left and bottom right) CMM; **46:** (top left and bottom) AC, (right) NMMB; **47:** (top) JC, (bottom) MM; **48:** (top) MM, (bottom) AC; **49:** NMMB; **50:** (both) GS; **51:** (both) GS; **52:** (both) GS; **53:** (both) GS; **54:** (top and center) JC, (bottom) GS; **55:** (top and right) GS, (left) JC; **56/57:** GS; **58:** (top) NMMB, (bottom) JC; **59:** (both) GS; **60:** (top left) JC, (top right) CM, (bottom) AC; **61:** (both) NMMB; **62:** JC; **63:** (top left, top right and bottom right) NMMB, (bottom left) H; **64:** (both) H; **65:** (top) Quadrant, (bottom) NMMB; **66:** (top) NMMB, (bottom) H; **67:** (top) NMMB, (bottom) H; **68:** (both) H; **69:** (top) NMMB, (bottom) Quadrant; **70/71:** H; **72:** (both) NMMB; **73:** (top) JC, (center) NMMB, (bottom) AC; **74:** (both) AC; **75:** (both) GS; **76:** (both) GS; **77:** (both): GS; **78:** (both) GS; **79:** (top) GS, (bottom) AC; **80/81:** GS; **82:** (all) NMMB; **83:** (top and center) NMMB, (bottom left and right) MM; **84:** (top) JC, (bottom) Quadrant; **85:** (top) K, (bottom) MM; **86:** (both) K; **87:** (both) K; **88:** (both) K; **89:** (top) Quadrant, (bottom) K; **90:** (top) CM, (bottom) K; **91:** (top) CM, (center and bottom) AC; **92:** NMMB; **93:** (both) CM; **94:** (top) AC, (bottom) MM; **95:** (top) MM, (bottom) Quadrant; **96:** (top) JC, (bottom) CM; **97:** (both) CM; **98:** (top) NMMB, (bottom) CM; **99:** NMMB; **100/101:** GS; **102:** (top left) NMMB, (top right and bottom left) CM, (bottom right) AC; **103:** (both) MM; **104:** (all) CM; **105:** (top) CM, (bottom) MM; **106:** (both) CM; **107:** (top) Quadrant, (bottom) MM; **108:** (top) AC, (bottom) MM; **109:** NMMB; **110:** (top) JC, (bottom) NMMB; **111:** (both) AC; **112:** (top) CM, (bottom) AC; **113:** (top) MM, (bottom) AC; **114/115:** GS; **116:** (top left and bottom) MM, (center) NMMB; **117:** (top and center) NMMB, (bottom) AC; **118:** MM; **119:** (all) NMMB; **120:** (top) NMMB, (bottom) Quadrant; **121:** (top) NMMB, (bottom left and bottom right) Quadrant; **122/123:** GS; **124:** (top) MM, (bottom) Quadrant; **125:** (top) GS, (bottom) Quadrant; **126:** MM; **127:** (top left) CM, (top right and bottom) MM; **128:** (top and left) CM, (bottom) MM; **129:** (top) MM, (bottom) NMMB; **130:** (both) AC; **131:** (top) NMMB, (bottom) CM; **132:** (top left) AC, (top right) CM, (bottom) GS; **133:** MM; **134/135:** GS; **136:** (top) AC, (bottom) GS; **137:** JC; **138:** (top left) JC, (top right and bottom) NMMB; **139:** (top) GS, (bottom) JC; **140:** (top) MM, (bottom) AC; **141:** (top) GS, (bottom) MM; **142/143:** NMMB; **144:** NMMB; **145:** (both) NMMB; **146:** MM; **147:** (both) NMMB; **148:** (both) MM; **149:** (top) MM, (bottom) AC; **150:** (top) NMMB, (bottom) AC; **151:** (both) MM; **152:** Quadrant; **153:** (top and bottom center) NMMB, (bottom left and bottom right) AC; **154/155:** AC; **156:** JC; **157:** (both) MM; **158:** AC; **159:** MM; **160:** (top) NMMB, (bottom) JC; **161:** (both) NMMB; **162:** (top) NMMB, (bottom) MM; **163:** (both) Quadrant; **164:** (top) MM, (center and bottom) AC; **165:** (both) MM; **166/167:** GS; **168:** (both) AC; **169:** NMMB; **170:** (top) AC, (bottom) NMMB; **171:** (top) GS, (bottom) NMMB; **172:** NMMB; **173:** (top) AC, (bottom) MM; **174:** (top) AC, (bottom) MM; **175:** (both) MM; **176/177:** GS; **178:** MM; **179:** (both) MM; **180:** (both) Y; **181:** (top) Quadrant, (left and right) NMMB; **182/183:** Quadrant; **184:** (top left) Quadrant, (top right) NMMB, (bottom) JC; **185:** (top) Y, (bottom) MM; **186:** (top) Y, (bottom) MM; **187:** MM.